R00219 78064

CHICAGO PUBLIC LIBRARY
HAROLD WASHINGTON LIBRARY CENTER

R0021978064

REF
QE
431.6
.M3
S72
cop. 1

FORM 125M

NATURAL SCIENCES
and USEFUL ARTS

Chicago Public Library

Received _____ JUN 1 7 1975 _____

D1609662

THE PHYSICAL PRINCIPLES OF ROCK MAGNETISM

FURTHER TITLES IN THIS SERIES

1. *F.A. VENING MEINESZ*
THE EARTH'S CRUST AND MANTLE

2. *T. RIKITAKE*
ELECTROMAGNETISM AND THE EARTH'S INTERIOR

3. *D.W. COLLINSON, K.M. CREER and S.K. RUNCORN*
METHODS IN PALAEOMAGNETISM

4. *M. BÅTH*
MATHEMATICAL ASPECTS OF SEISMOLOGY

Developments in Solid Earth Geophysics
5

THE PHYSICAL PRINCIPLES OF ROCK MAGNETISM

by

FRANK D. STACEY

Professor of Applied Physics
University of Queensland, Brisbane, Australia

and

SUBIR K. BANERJEE

Associate Professor of Geophysics
University of Minnesota, Minneapolis, U.S.A.

ELSEVIER SCIENTIFIC PUBLISHING COMPANY
Amsterdam — London — New York   1974

ELSEVIER SCIENTIFIC PUBLISHING COMPANY
335 JAN VAN GALENSTRAAT
P.O. BOX 211, AMSTERDAM, THE NETHERLANDS

AMERICAN ELSEVIER PUBLISHING COMPANY, INC.
52 VANDERBILT AVENUE
NEW YORK, NEW YORK 10017

REF
QE
431.6
.M3
S72
Cop. 1

LIBRARY OF CONGRESS CARD NUMBER: 72-87965

ISBN: 0-444-41084-8

WITH 72 ILLUSTRATIONS AND 13 TABLES

COPYRIGHT © 1974 BY ELSEVIER SCIENTIFIC PUBLISHING COMPANY, AMSTERDAM

ALL RIGHTS RESERVED. NO PART OF THIS PUBLICATION MAY BE REPRODUCED, STORED IN A RETRIEVAL SYSTEM, OR TRANSMITTED IN ANY FORM OR BY ANY MEANS, ELECTRONIC, MECHANICAL, PHOTOCOPYING, RECORDING, OR OTHERWISE, WITHOUT THE PRIOR WRITTEN PERMISSION OF THE PUBLISHER,
ELSEVIER SCIENTIFIC PUBLISHING COMPANY, JAN VAN GALENSTRAAT 335, AMSTERDAM

PRINTED IN THE NETHERLANDS

PREFACE

Since the early 1950's rock magnetism has become an important common interest for physicists and geologists. Important to geologists largely because it has revived the hypothesis of continental drift, which is now recognized as an intrinsic part, or even an incidental consequence, of global geological processes, and to many physicists because the need to study the physical problems of rock magnetization has provided the initial incentive to extend their interests into the rewarding field of geophysics. Our aim in this text is to present a review of the physical principles which is of modest length and meets the dual requirements of being comprehensible to geologists and satisfying to physicists.

Even a very small percentage of a ferromagnetic mineral, such as magnetite, in a rock masks the paramagnetic and diamagnetic properties of the other minerals. The properties which we measure are thus the properties of assemblies of small, and usually independent magnetic grains. The physical study of rock magnetism is therefore concerned with the properties of finely divided magnetic materials. Solid-state physicists have discovered the art of making stable permanent magnets by producing very fine grained materials. The fact that many rocks are stable permanent magnets (although weak ones) is therefore reasonably well understood. The geophysical significance of this simple discovery cannot be overemphasized. The fact that rocks have recorded faithfully the directions of the geomagnetic field at the times when they were formed, many millions of years ago, has provided a quantitative key to the early history of the earth which ranks in importance with the recognition that abundances of isotopes in radioactive decay schemes allow absolute ages of rocks to be determined.

It should be admitted that the physical study of rock magnetism has followed the development of paleomagnetism, which, although using certain physical techniques, is based on essentially geological knowledge. The rapid development of paleomagnetism left a series of questions and doubts concerning the origin and stability of rock magnetizations, which physicists have been struggling to answer. This book records the measure of their success. It also indicates a fall-out of related discoveries: the recognition of the importance of remanent magnetism in the interpretation of magnetic surveys, the use of magnetic anisotropy as an indicator of rock fabric and of the possibility that significant piezomagnetic changes accompany seismic activity.

The subject has developed some of its own jargon, which conflicts in a minor way with conventional physical usage. Thus, it is convenient to refer to the slight magnetizations of rocks acquired naturally or in small laboratory fields as *remanence*, distinguishing the special case of remanence acquired by exposure to a very high field as *saturation remanence*. Generally, however, no confusion arises as we distinguish the remanence acquired in various ways by special terms, thermoremanent magnetization or TRM (due to cooling in a field), anhysteretic remanent magnetization or ARM (small steady field with superimposed decreasing alternating field), detrital remanent magnetization or DRM (magnetic grain alignment during sedimentary deposition).

To scientists brought up on rationalized M.K.S. units, a word in defence of our exclusive use of the C.G.S. electromagnetic units may be needed. Until now the M.K.S. system has suffered two major disadvantages. Firstly in treating magnetism it was usually presented in a manner which implied that B not H was fundamental. Fortunately this fallacy has now been refuted by experimental evidence (Whitworth and Stopes-Roe, 1971); this will allow the M.K.S. treatment to be developed as a rational as well as rationalized system, in which we do not have the possibility of omitting μ_0 from the value of magnetic moment, as is still common in the majority of elementary texts. However, even when this correction has filtered through the teaching system, there remains the problem that the electromagnetic units are the practical units of our subject; the earth's field has a strength of order 1 oersted (Oe) and the magnetization of magnetite in this field is close to 1 e.m.u.

We regard this book as a supplement to Irving's (1964) treatise on paleomagnetism, which states the essential geological problems and reviews the data obtained to 1963. Our approach is different from that adopted by Nagata (1953, 1961), in that we are less concerned with the conventional measuring techniques in rock magnetism, which are comprehensively discussed in the conference volume edited by Collinson et al. (1967); Nagata's (1953) first edition remains a useful introduction to the subject. Although in many respects outdated, reviews by Néel (1955) and Stacey (1963) are the principal sources of the ideas presented in the theoretical parts of this book.

The data in Appendixes 1–4 were prepared for this work by J.W. Higbie. We also appreciate the readiness of numerous authors to agree to our reproduction of their diagrams, of which the sources are indicated in captions. And we extend thanks to L.G. Parry and D.E.W. Gillingham, who scrutinized the manuscript, and to Mrs. Sue Clarke who maintained the typescript through numerous revisions.

CONTENTS

PREFACE V

CHAPTER 1. MAGNETIC PROPERTIES OF SOLIDS 1
  1.1 Introduction 1
  1.2 The spontaneous magnetization of ferromagnetics 3
  1.3 Exchange interactions 10
  1.4 Antiferromagnetism and ferrimagnetism 14
  1.5 The origin of anisotropy and magnetostriction 22

CHAPTER 2. MAGNETIC MINERALS 25
  2.1 Ferrimagnetism of magnetite 25
  2.2 Properties of titanomagnetites 28
  2.3 Maghemite 30
  2.4 Oxidized titanomagnetites 32
  2.5 Weak ferromagnetism in hematite 35
  2.6 Titanohematites 37
  2.7 Minor magnetic minerals 39

CHAPTER 3. FERROMAGNETIC DOMAIN THEORY 41
  3.1 Magnetostatic energy 41
  3.2 Anisotropy energy 47
  3.3 Magnetostrictive strain energy 49
  3.4 Domain walls 53
  3.5 Single domains and multidomains 58
  3.6 Domain wall moments and Barkhausen discreteness 60
  3.7 Lamellar intergrowths 63

CHAPTER 4. PROPERTIES OF MAGNETITE GRAINS AND OF ROCKS CONTAINING THEM 66
  4.1 Coercive force and grain size 66
  4.2 Susceptibility of magnetite grains 70
  4.3 Anisotropy in magnetite-bearing rocks 75
  4.4 Saturation remanence and coercivity of remanence 79
  4.5 Deflection of magnetization in a strongly magnetic layer 83

CHAPTER 5. PROPERTIES OF HEMATITE GRAINS 85
  5.1 Anisotropy and coercivity 85

5.2	Susceptibility of hematite	87
5.3	Anisotropy and rotational hysteresis in hematite-bearing rocks	89
5.4	Titanohematites	91
5.5	The maghemite-to-hematite transition	92

CHAPTER 6. THERMAL ACTIVATION EFFECTS 96
6.1	The concept of blocking temperature	96
6.2	Magnetic viscosity and paleomagnetic stability	98
6.3	Superparamagnetism and superantiferromagnetism	102

CHAPTER 7. THERMOREMANENT MAGNETIZATION (TRM) 105
7.1	Introduction	105
7.2	Néel's (1955) theory of TRM in single domain grains	105
7.3	TRM in large multidomains	107
7.4	The pseudo single domain (grain surface) effect	110
7.5	Total thermoremanence	113
7.6	The Koenigsberger ratios	115
7.7	Partial thermoremanences and the law of additivity	116
7.8	Thermoremanence in anisotropic rock	118

CHAPTER 8. DEPOSITIONAL REMANENT MAGNETIZATION (DRM) . . . 121
8.1	The occurrence of DRM	121
8.2	The detrital magnetization process	122
8.3	Demagnetization and stability of detrital remanence	125
8.4	Inclination errors, water flow and other extraneous effects	126

CHAPTER 9. CHEMICAL REMANENT MAGNETIZATION (CRM) 128
9.1	The paleomagnetic significance of CRM	128
9.2	The process of chemical magnetization	130
9.3	The stability of chemical remanence	134

CHAPTER 10. ALTERNATING FIELD DEMAGNETIZATION AND ANHYSTERETIC MAGNETIZATION 136
10.1	The demagnetization method	136
10.2	Magnetostatic forces in the demagnetization process	137
10.3	Anhysteretic remanent magnetization (ARM)	141

CHAPTER 11. PIEZOMAGNETIC EFFECTS 146
11.1	Effect of stress on susceptibility	146
11.2	Effects of stress upon remanence	150
11.3	Magnetostriction and paleomagnetism	155
11.4	The seismomagnetic and volcanomagnetic effects	158

CONTENTS ix

CHAPTER 12. REVERSALS OF REMANENT MAGNETIZATION 162
   12.1 Evidence for reversals of the geomagnetic field 162
   12.2 Self-reversal mechanisms 165

CHAPTER 13. MAGNETISM IN METEORITES 170
   13.1 Meteorite types and meteoritic iron 170
   13.2 Natural remanence in chondritic meteorites 172

APPENDIX 1. Table of the function $F'(a) = \int_0^1 x \tanh(ax)\, dx$ 176

APPENDIX 2. Table of the function $F(a) = -\int_0^1 \int_0^1 xy \tanh(axy)\, dx\, dy$ . . . 176

APPENDIX 3. Table of the Langevin Function, $L(a) = \coth(a) - 1/a$ 176

APPENDIX 4. Table of the function $F''(a) = \frac{1}{a} \ln\left(\frac{\sinh a}{a}\right)$ 177

BIBLIOGRAPHY 178

NAME INDEX 187

SUBJECT INDEX 191

Chapter 1

MAGNETIC PROPERTIES OF SOLIDS

1.1 INTRODUCTION

We are concerned with the properties of mineral grains which are *ferromagnetic*, i.e., iron-like or strongly magnetic, even though they normally occur in low concentrations, so that the rocks containing them may be only very weakly magnetic. Ferromagnetism is a special case of *paramagnetism*, which is the property of a material drawn towards the stronger part of a magnetic field. The opposite behaviour, repulsion by a magnetic field, is *diamagnetism*. Although P. Langevin gave quantitative classical explanations of both paramagnetism and diamagnetism, in terms of the responses to applied magnetic fields of electrons in atomic orbits, a rigorous classical treatment of electron orbits shows these opposite effects to be exactly mutually cancelling, so that no magnetic properties of matter appear at all (Miss Van Leeuwen's Theorem — see Van Vleck, 1932, pp.94–100). The magnetic properties thus depend essentially on the quantized nature of matter.

A quantum-mechanical treatment of the fundamental magnetic behaviour of matter is beyond the scope of this book, but the features to which we refer in later chapters will be discussed in an elementary way in sections *1.3–1.5*. This is necessary to an understanding of some of the newer developments of the subject, which appeal to the atomic nature of magnetism.

Macroscopic magnetic properties are described in terms of ferromagnetic domain theory (Chapter 3) for which the existence of ferro- (or ferri-)magnetism can be assumed as a starting point.

The basic equations expressing the magnetization of a material relate its magnetic moment per unit volume I to the field H and to the magnetic flux density B. In the electromagnetic system of units, which is used in this book:

$$B = \mu H = H + 4\pi I \quad (1.1)$$

$$I = \chi H \quad (1.2)$$

$$\mu = 1 + 4\pi\chi \quad (1.3)$$

where μ is *permeability*, and χ is *susceptibility*. The equivalent equations in the rationalized M.K.S. system are:

$$B = \mu H = \mu_0 H + I$$

$$I = \chi H$$

$$\mu = \mu_0 + \chi$$

where μ_0 is the permeability of free space ($4\pi \cdot 10^{-7}$ henry). In the M.K.S. system, care must be taken to avoid the ambiguity in units of magnetic moment ($I \times$ volume). Different texts use either weber-meters, as follows from above equations, or ampere-turn m^2; these units differ by the factor μ_0.

The magnetic moment of any body is associated with circulating electric charges within it. Except for the special problem of nuclear magnetic resonance, the effect of electron motion is completely dominant, so that the gyromagnetic ratio of mechanical angular momentum to magnetic moment is a characteristic of the charge to mass ratio of electrons, e/m. The value of the gyromagnetic ratio for orbital electron motion, in electromagnetic units[1], is:

$$\rho_0 = 2m/e \tag{1.4}$$

Now angular momentum is quantized. It cannot have any arbitrary value, only a multiple of $\hbar = h/2\pi$, where h is Planck's constant. Therefore magnetic moments are quantized in units of:

$$\mu_B = \hbar/\rho_0 = 9.27 \cdot 10^{-21} \text{ erg oersted}^{-1} \tag{1.5}$$

This unit is the Bohr magneton. It is the magnetic moment associated with the orbital motion of an electron in the first Bohr orbit of atomic hydrogen and is the fundamental unit of magnetic moment.

In addition to their orbital motions, electrons also have an intrinsic angular momentum or spin. The spin angular momentum is $\hbar/2$, i.e., half of the smallest unit of orbital angular momentum, but the magnetic moment of electron spin is still 1 Bohr magneton ($1\mu_B$)[2]. Thus the gyromagnetic ratio of electron spin, ρ_S, is half the orbital value, i.e.:

$$\rho_S = m/e \tag{1.6}$$

This property of electron spin is in conflict with the intuition of classical physics; its explanation is given by relativistic quantum mechanics developed by P.A.M. Dirac. The spin angular momentum of an electron may be $\pm \hbar/2$ but not zero, i.e., only changes by $\pm \hbar$ are allowed, as in orbital transitions; correspondingly the magnetic moment is $\pm 1\mu_B$, but cannot be zero.

Gyromagnetic ratios are conveniently represented by a simple numerical factor g':

$$\rho = (1/g')(2m/e) \tag{1.7}$$

For orbital motion $g' = 1$ and for spin $g' = 2$. Measurements of g' for any material thus indicate immediately the ultimate nature of its magnetic moments. The most reliable measurements are those made by the Richardson-Einstein-De Haas effect, whereby sud-

---

[1] In M. K. S. units multiply by μ_0. The inverse ratio is known as the magnetomechanical factor.
[2] According to the "vector model" of the atom the electron has an intrinsic spin moment of $\sqrt{3}\,\mu_B$ but precesses, so that the component in any measured direction is $1\mu_B$ (see Brailsford, 1966, p.42). However, for the present purpose we take the intrinsic spin moment as $1\mu_B$.

den magnetization of a body gives it an observable rotation, due to the transfer of angular momentum to the electrons in the magnetization process. Measurements on ferromagnetics all indicate g' very close to 2, i.e., a complete dominance of spin magnetic moments. The orbital magnetic moments are said to be *quenched*, that is atomic interactions cause them to pair-off, giving almost complete mutual cancellation.

In common with the familiar ferromagnetic metals, the properties of magnetic minerals are due to spin-controlled interactions between electrons. These are exchange interactions, which cause mutual alignment of spins over a large number of atoms in spontaneous magnetization, whose temperature-dependence is considered in the following section. Discussion of the interactions themselves is deferred to section *1.3*.

1.2 THE SPONTANEOUS MAGNETIZATION OF FERROMAGNETICS

The theory of ferromagnetism hangs on the existence of exchange interactions (section *1.3*) but experimentally ferromagnetics are characterized by hysteresis, or irreversibility of magnetization (Fig.1.1). The magnetic polarization of a ferromagnetic

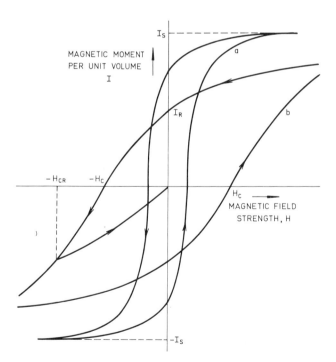

Fig.1.1. Hysteresis loops typical of (*a*) soft, and (*b*) hard magnetic materials, showing, for curve *b*, the quantities defined as saturation remanence I_R, coercive force H_C and coercivity of remanence H_{cR}, and for curve *a* the saturation magnetization I_S. Note that this indicates intrinsic properties of the magnetic material. Observed properties of dispersed grains are shown in Fig.4.5.

material depends upon its magnetic history and not merely upon the field to which it is exposed at a particular instant, so that it can remain magnetized after removal of a field. We are here interested in the properties of so-called *hard* magnetic materials, which resist demagnetization well, as illustrated by curve b in the figure. Weiss(1907) first clearly stated that the irreversibility had its origin in the magnetic subdivision of a ferromagnetic body into *domains*, each of which is spontaneously magnetized to saturation by virtue of a strong molecular field. The direction of domain magnetization is determined by crystallographic structure, but in all except very fine grains the domains are mutually oriented so that the total magnetization of a ferromagnetic body may have any value between zero and saturation, depending upon the fields to which it is exposed and was previously exposed.

The spontaneous magnetization of the domains is a function of temperature and disappears at the *Curie point* (580°C for magnetite). At this temperature thermal agitation overcomes the interactions which produce mutual alignment of the atomic magnetic moments. At 0°K the alignment of moments is complete and at intermediate temperatures there is some misalignment by thermal agitation. The aligning force is an exchange interaction whereby the mutual electrostatic energy of two atoms is lower if the unbalanced magnetic moments of the electrons in their tightly bound 3d shells are parallel to one another. The probabilities of the moment of a particular atom being either parallel or antiparallel to the moments of its neighbours are therefore determined by the orientations of the neighbouring moments and the net magnetic alignment of a whole assembly is a complex (and still unsolved) statistical problem involving all of the mutual interactions. This is known as the Ising problem. However, even if it were solved it would be of limited application because the simple quantum condition that an individual electron's magnetic moment must be either parallel or antiparallel to an externally applied field is compromised in an assembly, in which it is only the total magnetic misalignment which is quantized.

The essential features of atomic alignment in ferro- and ferri-magnetism are explained on the hypothesis of a *molecular field*, as proposed by Weiss(1907). Weiss's work antedates by many years the quantum theory upon which explanations of magnetic properties are now based. What follows is an elementary treatment of the quantum-corrected Weiss theory.

Suppose that a particular atom in a magnetic lattice has z nearest neighbours and that with each of them there is an exchange interaction such that if the magnetic moments of a particular pair are parallel their mutual energy is $-A$ and if they are antiparallel the energy is $+A$. The hypothesis that these interactions are equivalent to an intermolecular field, characteristic of the whole material, acting on the central atom, is equivalent to supposing that the z nearest neighbours have an average alignment which is the same as the average of all atoms in the material; local statistical variations are neglected. In a body-centered cubic lattice $z = 8$ and in face-centred cubic or hexagonal close-packed lattices $z = 12$, and these numbers are just sufficient to make the intermolecular field a reasonable approximation.

Now suppose that, of the z nearest neighbours, p have moments aligned parallel to some preferred direction and a are antiparallel, so that:

$$p + a = z \tag{1.8}$$

and since these z neighbours are representative of all of the atoms in the material, the spontaneous magnetization I_S (magnetic moment per unit volume), relative to the magnetization I_{S0} by complete alignment at $0°K$ is:

$$I_S/I_{S0} = (p - a)/(p + a) \tag{1.9}$$

so that:

$$p = \frac{z}{2}\left(1 + \frac{I_S}{I_{S0}}\right) \tag{1.10}$$

$$a = \frac{z}{2}\left(1 - \frac{I_S}{I_{S0}}\right) \tag{1.11}$$

But the energy of the considered central atom, relative to the demagnetized state ($p = a = z/2$; $I_S/I_{S_0} = 0$) is:

$$E = \mp (p - a) A = \mp zA \frac{I_S}{I_{S0}} \tag{1.12}$$

depending upon whether its magnetic moment is parallel or antiparallel to the preferred direction. If the atomic magnetic moment is μ and the effective inter-molecular field is F, then also:

$$E = \mp F\mu = \mp \lambda I_S \mu \tag{1.13}$$

where λ is the intermolecular field coefficient. By comparing (1.12) and (1.13):

$$\lambda = zA/\mu I_{S0} \tag{1.14}$$

The atomic magnetic moments experience an aligning force which is proportional to the alignment itself. This is an example of a cooperative phenomenon. Above a critical temperature (the Curie point) the atomic moments are thermally disordered[1], but, on cooling below the Curie point θ, order rapidly sets in, so that at 0.9θ the magnetic alignment is already 60% complete.

Now consider the response of independent (uncoupled) electron spins, of moments μ, to a field H. They may be either parallel or antiparallel to the field and their energies

---

[1] Short range order, i.e., the tendency for parallelism of atomic moments which are near one another in the lattice, persists above the Curie point; it is long-range order, the parallelism over many lattice spacings which is responsible for ferromagnetism and which disappears at the Curie point.

are then $\mp \mu H$, i.e., the parallel state is favoured by an energy difference $2\,\mu H$. The parallel state is more probable by the factor $e^{2\mu H/kT}$, where k is Boltzmann's constant. But the sum of the probabilities of being parallel (P_+) or antiparallel (P_-) is unity so that:

$$P_+ = e^{2\mu H/kT}/(e^{2\mu H/kT} + 1) \tag{1.15}$$

$$P_- = 1/(e^{2\mu H/kT} + 1) \tag{1.16}$$

and for the whole assembly:

$$\frac{I_S}{I_{S0}} = \frac{P_+ - P_-}{P_+ + P_-} = \tanh\left(\frac{\mu H}{kT}\right) \tag{1.17}$$

In the general problem of magnetic alignment of atomic moments μ with angular momenta $Jh/2\pi$ (h = Planck's constant), there are $(2J + 1)$ discrete values for the orientation of a moment in the field and we obtain the result:

$$\frac{I_S}{I_{S0}} = B_J\left(\frac{\mu H}{kT}\right) = \frac{2J+1}{2J}\coth\left(\frac{2J+1}{2J}\cdot\frac{\mu H}{kT}\right) - \frac{1}{2J}\coth\left(\frac{1}{2J}\cdot\frac{\mu H}{kT}\right) \tag{1.18}$$

B_J is known as the Brillouin function and eq. 1.17 is the special case $J = 1/2$, appropriate to independent electron spins. For other small values of J, the form of eq. 1.18 is very similar to 1.17.

In the study of paramagnetic materials a combination of very low temperatures and very high fields must be used to obtain the condition $\mu H > kT$, so that the form of eq. 1.17 or 1.18 can be verified. In normal laboratory experience $\mu H/kT \ll 1$ and magnetization I is linear in field H:

$$\frac{I}{I_{S0}} = \frac{J+1}{3J}\cdot\frac{\mu H}{kT} \tag{1.19}$$

so that the susceptibility:

$$\chi = \frac{I}{H} = \frac{J+1}{3J}\cdot\frac{\mu I_{S0}}{kT} = \frac{J+1}{3J}\cdot\frac{N\mu^2}{kT} \tag{1.20}$$

where N is the number of atomic moments per unit volume. This is Curie's law, obeyed by paramagnetic materials in which the atomic moments are sufficiently widely separated not to interact with one another.

To explain ferromagnetism, we must add to the field H in eq. 1.17 or 1.18 an intermolecular field λI (eq. 1.13), which may be very much stronger than H. Using eq. 1.17:

$$\frac{I}{I_{S0}} = \tanh\left[\frac{\mu}{kT}(H + \lambda I)\right] \tag{1.21}$$

There are two alternative conditions to consider: (a) I very small, so that $I/I_S \ll 1$; and (b) I large, so that $\lambda I \gg H$. Considering first condition (a), the function tanh is equal to its argument and we can rearrange terms to obtain:

$$\chi = \frac{I}{H} = \frac{\mu I_{S0}/kT}{1 - \lambda\mu I_{S0}/kT} = \frac{C}{T-\theta} \qquad (1.22)$$

where $C = \mu I_{S0}/k = N\mu^2/k$ is known as the Curie constant and $\theta = \lambda C$ is the Curie temperature of the material. This is the Curie-Weiss law of paramagnetism, a generalization of the Curie law (eq. 1.20) which accounts for the molecular interactions. Eq. 1.22 represents the paramagnetic susceptibility of a ferromagnetic material above its Curie point, θ (Fig. 1.2). As $T \to \theta$, so $\chi \to \infty$ and the condition $I/I_{S0} \ll 1$, used to abtain eq. 1.22, breaks down. By considering the alternative condition (b) we may show that θ is the temperature at which spontaneous magnetization begins to appear.

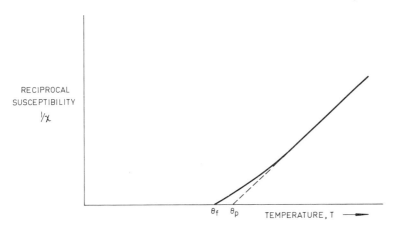

Fig.1.2. The paramagnetic susceptibility of a ferromagnetic above its Curie point. The linear relationship between $1/\chi$ and T is the Curie-Weiss law (eq.1.22) and the intercept is the paramagnetic Curie temperature θ_p. This differs slightly (by 15°C in the well documented case of nickel) from the ferromagnetic Curie point θ_f, which marks the onset of spontaneous magnetization. The curvature of the graph just above θ_f, was explained by Néel (1940) in terms of the persistence of short-range magnetic order above θ_f.

Neglecting H in eq. 1.21 we can re-write the equation to give the spontaneous magnetization I_S:

$$\frac{I_S}{I_{S0}} = \tanh\left(\frac{\mu\lambda I_{S0}}{kT}\frac{I_S}{I_{S0}}\right) = \tanh\left(\frac{\theta}{T}\frac{I_S}{I_{S0}}\right) \qquad (1.23)$$

since θ is the critical (Curie) temperature by eq. 1.22. For $T > \theta$ there is no solution to (1.23) except $I_S = 0$, i.e., there is no spontaneous magnetization above this temperature. For $T < \theta$, solutions of (1.23) give $I_S/I_{S0} > 0$. Values may be obtained numerically, or graphically as in Fig. 1.3. The resulting spontaneous magnetization curve has the form of the experimentally observed curves; Fig. 2.1 shows the spontaneous magnetizations of the two sub-lattices in magnetite.

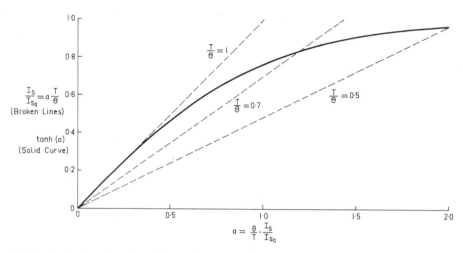

Fig.1.3. Graphical solution of eq.1.23 to obtain the curve of spontaneous magnetization vs temperature. Using the parameter $a = (\theta/T) \times (I_S/I_{S0})$, the value of I_S/I_{S0} at a particular value of T/θ is obtained from the intersection of graphs of $\tanh(a)$ and $a \cdot T/\theta = I_S/I_{S0}$. The observed spontaneous magnetization curve of magnetite is indicated for two sub-lattices separately in Fig.2.1.

There is a satisfactory general agreement of this elementary theory with observations on the ferromagnetic metals and also on magnetite. The most serious discrepancy is at low temperatures and is not apparent in Fig. 2.1. Eq. 1.23 gives an exponential approach of I_S/I_{S0} to absolute saturation as $T/\theta \to 0$, whereas observations show:

$$\left(1 - \frac{I_S}{I_{S0}}\right) \propto T^{3/2} \tag{1.24}$$

at low temperatures. This is indicative of a basic defect in the theory, which assumes that the magnetic misalignment is due to the reversals of individual atomic moments. In fact the misalignment is due to *spin waves*, which may be visualized as waves in the array of magnetic moments in a solid, similar to the waves in a field of wheat across which a wind is blowing. The energy of a wave in the spin structure, in which the angles between spin orientations of neighbouring atoms are very small, but the total magnetic misalignment is $2\mu_B$, is very much less than the energy of reversal of a single electron spin. Spin waves are therefore more easily excited than are individual spin reversals[1] and account for the deviation from absolute saturation at low temperatures.

In rock magnetism our interest in very low temperature properties is restricted to two effects, the processes of demagnetization which accompany phase changes in magnetite and hematite, and the low-temperature evidence for cation distribution between non-equivalent lattice sites in ferrites. However, spin waves themselves are important.

---

[1] For a discussion of spin waves with illustrations of the motion, see Kittel (1971), p. 538.

They provide the mechanism whereby thermal excitation affects the magnetic properties of ferromagnetic materials (Chapter 6). The structure of spin waves is closely related to that of domain walls (section 3.3).

Direct evidence of the energy involved in exchange interactions is provided by the specific heat anomaly at the Curie point. In terms of the molecular field, the magnetic energy of a material with spontaneous magnetization I_S is:

$$E = -\int_0^{I_S} F\,dI = -\int_0^{I_S} \lambda I\,dI = -\frac{1}{2}\lambda I_S^2 \tag{1.25}$$

But the magnetic contribution to specific heat is:

$$C_M = \frac{dE}{dT} = -\frac{1}{2}\lambda \frac{d}{dT}(I_S^2) \tag{1.26}$$

From the variation of (I_S/I_{S0}) with T, we therefore obtain the curve of C_M vs T which appears as a λ-point anomaly superimposed on the specific heat curve, as in Fig. 1.4. The

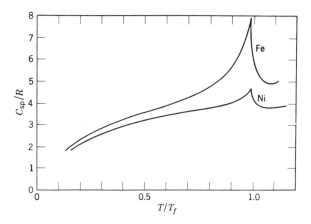

Fig. 1.4. Specific heat anomalies associated with the magnetic order–disorder transitions at the Curie points θ_f of ferromagnetic materials. The specific heats are given in cal./mole and temperatures are normalised, $T_f \equiv \theta_f$ being the Curie point. The magnetic contribution to specific heat extends above θ_f by virtue of the persistence of short-range magnetic order above θ_f. This is the same effect as produces the curvature of the $1/\chi$–T curve in Fig. 1.2.
(Figure reproduced by permission from Morrish, 1965.)

magnitude of the peak in C_M can be calculated by expanding the hyperbolic tangent in (1.23) to the term in $(I_S/I_{S0})^3$, since (I_S/I_{S0}) is small at $T \to \theta$:

$$\frac{I_S}{I_{S0}} = \frac{I_S}{I_{S0}}\frac{\theta}{T} - \frac{1}{3}\left(\frac{I_S}{I_{S0}}\frac{\theta}{T}\right)^3 + \ldots \tag{1.27}$$

so that:

$$\frac{d}{dT}(I_S{}^2) = 3I_{S0}{}^2 \frac{d}{dT}\left\{\left(\frac{T}{\theta}\right)^2 - \left(\frac{T}{\theta}\right)^3\right\} = -3I_{S0}{}^2/\theta \quad \text{at } T \to \theta \quad (1.28)$$

But, as in eq. 1.22, $\theta = \lambda I_{S0}{}^2/Nk$, so we obtain:

$$(C_M)_{\max} = \frac{3}{2}Nk = 3 \text{ cal. mole}^{-1}\,°C^{-1}. \quad (1.29)$$

This is quite close to the observed magnitude in most materials. It is to be noted that antiferromagnetics and ferrimagnetics all have similar λ-point specific heat anomalies at their transition temperatures.

1.3   EXCHANGE INTERACTIONS

The magnetostatic interaction between the spin moment of an electron and the moments of single electrons on its z neighbouring atoms in a solid of atomic spacing a, corresponds to a molecular field at saturation:

$$H_M = \frac{z\mu_B}{a^3} \approx 5000 \text{ oersteds} \quad (1.30)$$

for $z = 8$. The corresponding Curie temperature θ_M would then be given by:

$$\theta_M \approx \frac{\mu_B H_M}{k} = 0.3°K \quad (1.31)$$

where k is Boltzmann's constant. Thus if its existence depended upon magnetostatic interactions, ferromagnetism would be observed only at temperatures lower than a few tenths of 1°K. Observed Curie points are several hundred degrees Kelvin, the highest being 1400°K for cobalt, so that the spin alignment is several thousand times stronger than that provided by magnetic dipole interaction. Electrostatic interactions have energies in the appropriate range, i.e., comparable to molecular binding energies. The ferromagnetic interactions, known as exchange interactions, are spin-dependent electrostatic interactions between electrons on neighbouring atoms, as first recognized by W. Heisenberg. The spin-dependence makes them similar to magnetic interactions, although much stronger. The homopolar chemical bond is another example of spin-dependent electrostatic interaction and the same general theory suffices for both. A simplified presentation, following closely the treatment by Heitler (1945), is given here.

We consider two atomic nucleii a and b, with two orbital electrons 1 and 2, and suppose initially that electron 1 is in orbit about nucleus a and 2 about b. The electrons are not point charges but must be thought of as clouds of charge about each of the nucleii. The charge distribution (which may also be regarded as the distribution of probability of finding an electron at a particular place) is then represented by a *wave function*. For the two electrons we have two wave functions $\psi_a(1)$ and $\psi_b(2)$. These are geometrical functions whose squares represent the charge or probability distributions of the two electrons in the spaces about the two nucleii. We can therefore take the product of these functions to represent the simultaneous probabilities (per unit volume) of finding 1 in an elementary volume near a and 2 in a volume near b. A wave function for the two electrons is therefore $\psi_a(1) \cdot \psi_b(2)$. Now it is a property of electrons that they are fundamentally indistinguishable from one another. Thus an equally acceptable wave function is obtained by *exchanging* 1 and 2: $\psi_a(2) \cdot \psi_b(1)$. But neither of these wave functions alone is adequate because it associates each electron with a particular nucleus. We must therefore take linear combinations of these wave functions which give equal weights to the association of each electron with each nucleus:

$$\psi_s = \psi_a(1) \cdot \psi_b(2) + \psi_a(2) \cdot \psi_b(1) \tag{1.32}$$

$$\psi_{as} = \psi_a(1) \cdot \psi_b(2) - \psi_a(2) \cdot \psi_b(1) \tag{1.33}$$

Here ψ_s is a symmetric wave function, that is, if 1 and 2 are interchanged the function remains the same. ψ_{as} is an antisymmetric wave function, that is interchanging 1 and 2 reverses its sign. However, note that the charge distribution represented by ψ_{as} is given by ψ_{as}^2 and is not affected by the exchange.

These wave functions are orbital wave functions, that is they represent the geometrical distribution of charges. The conditions of the electrons are not completely specified by orbital wave functions, which give no information about spin directions. We therefore consider additional *spin-wave functions* of simple form, α representing spin "up" and β spin "down", so that if both spins are "up" the spin wave function for the two electrons is $\alpha(1) \cdot \alpha(2)$, but if 2 is "down" then the function would be $\alpha(1) \cdot \beta(2)$. Again the indistinguishability compels us to take linear combinations of functions of this latter form to avoid identifying individual electrons with particular states. Four possible spin wave functions result:

	spin wave function	*"up" component of magnetic moment*
(symmetric)	$\alpha(1) \cdot \alpha(2)$	$+2\,\mu_B$
	$\beta(1) \cdot \beta(2)$	$-2\,\mu_B$
	$\alpha(1)\beta(2) + \alpha(2)\beta(1)$	0
(antisymmetric)	$\alpha(1)\beta(2) - \alpha(2)\beta(1)$	0

The complete wave function for the two electrons is obtained by multiplying an orbital wave function by a spin wave function, but only half of the possible combinations are allowed. Electrons are antisymmetric particles or *fermions*, which means that the wave function for a collection of electrons is antisymmetric (reverses sign) with respect to the exchange of any two. The complete wave function for the two electrons is antisymmetric only if the orbital wave function is symmetric and the spin wave function antisymmetric or vice-versa. The possible wave functions are therefore:

$$\psi_n = [\psi_a(1) \cdot \psi_b(2) + \psi_a(2) \cdot \psi_b(1)] \cdot [\alpha(1) \cdot \beta(2) - \alpha(2)\beta(1)] \quad (1.34)$$

$$\psi_p = [\psi_a(1) \cdot \psi_b(2) - \psi_a(2) \cdot \psi_b(1)] \cdot \begin{bmatrix} \alpha(1) \cdot \alpha(2) \\ \text{or} \\ \beta(1) \cdot \beta(2) \\ \text{or} \\ \alpha(1) \cdot \beta(2) + \alpha(2) \cdot \beta(1) \end{bmatrix} \quad (1.35)$$

ψ_n is the wave function for the two electrons with antiparallel spins, which have no total spin magnetic moment. ψ_p is the wave function for parallel spins with a moment of $2\,\mu_B$, which can have components $\pm 2\,\mu_B$ or zero in any direction distinguished by the environment (as by an external field). Eq. 1.34 and 1.35 express the essential point of the theory of exchange interactions. The geometrical charge distributions are different for the parallel and antiparallel spin configurations. The mutual potential energy due to charge overlap of the two electrons is thus dependent upon their relative spin orientations.

An approximate calculation of the energies of the two orbital states represented by ψ_s and ψ_{as} may be made by a perturbation method, assuming initially unperturbed single atomic functions for ψ_a, ψ_b and then evaluating the integral:

$$E = \iint V \psi^2 d\tau_1 d\tau_2 \quad (1.36)$$

where $d\tau_1, d\tau_2$ are elements of volume in the combined charge clouds of the two electrons, V is the coulomb potential energy for electrons situated at $d\tau_1$ and $d\tau_2$ and ψ^2 is either ψ_s^2 or ψ_{as}^2 and is the product of the charge concentrations in the two volumes. The energies of the two states are expressed in terms of geometrical integrals C, A, S:

$$E_n = \frac{C+A}{1+S} \quad (1.37)$$

$$E_p = \frac{C-A}{1-S} \quad (1.38)$$

where $S \ll 1$ and may be neglected in a simple discussion, so that the energy difference between parallel and antiparallel spin states is $2A$. A is the exchange integral or interaction energy, as used in (1.12). If the geometry is such that A is positive, then we say that we have positive exchange interaction, in which, by (1.37) and (1.38), parallel spins give lower energy than antiparallel spins. This is the interaction which causes parallel alignment of all neighbouring spins and thus leads to ferromagnetism. Most exchange interactions are negative (i.e. A is negative), as in the homopolar chemical bond, e.g., two hydrogen atoms which combine with their electron spins antiparallel. In magnetic phenomena negative exchange causes antiferromagnetism (and ferrimagnetism), in which magnetic moments of neighbouring atoms are aligned antiparallel.

The conditions which lead to positive exchange interaction have been much discussed since the original paper on this subject by W. Heisenberg, who pointed out that it was favoured by modest overlap of highly eccentric orbital wave functions. This condition occurs in elements of the first transition series (iron group) of the Periodic Table, in which the 4s electrons are responsible for metallic bonding and become the conduction electrons, virtually free to migrate, leaving ion cores with incomplete 3d shells. Exchange interactions between 3d electrons are responsible for ferromagnetism in these elements.

By the foregoing perturbation theory we can also consider the spins of several orbital electrons with a common nucleus, referring specifically to electrons in 3d states. There are 5 orbital wave functions, each associated with one spin "up" and one spin "down" state, so that the total occupation of the 3d shell is 10 electrons. Considering the interaction between two electrons, if their orbital states are the same ($\psi_a \equiv \psi_b$) then the antisymmetric orbital function vanishes. Only ψ_n is allowed, requiring the spins to be antiparallel, and no more than two electrons can have the same orbital state. This is *Pauli's exclusion principle*, which has been written into eq. 1.34 and 1.35 by restricting these equations to those which are antisymmetric with respect to the exchange of two electrons. Now if the orbital states are different we can carry out a perturbation calculation which shows that they have lower energy if their spins are parallel. This is *Hund's rule* of highest multiplicity, according to which electrons in a single atom have parallel spins, so far as Pauli's exclusion principle allows. This means that as the 3d states are filled progressively through the first transition series of elements, electrons are added with spins parallel until all five spin "up"/"down" states are occupied and only then do the spin "down"/"up" states start being occupied.

Hund's rule explains the importance of 3d electrons in magnetic phenomena. In rock magnetism we are concerned with the 3d electrons in iron, in which there are 8 electrons shared by the 4s (valence) states and 3d (magnetic) states. The ferrous ion, Fe^{2+}, uses two electrons in chemical bonding (i.e., in valence states) leaving six 3d electrons. By Hund's rule five are mutually parallel and the sixth antiparallel, giving a magnetic moment of 4 μ_B to Fe^{2+}. Similarly the ferric ion, Fe^{3+}, has only five 3d electrons, all parallel, giving it a magnetic moment of 5 μ_B.

In solids we are concerned not with isolated interactions between individual pairs of electrons but with very large numbers and this confuses the simple spectroscopic rule by which the components of spin-angular momentum S may have ($2S + 1$) values ($S = 1/2$ for a single electron and there are two states, spin "up" and "down"). The angle between any pair of spin axes may have any arbitrarily small value, provided the spin angular momentum of the whole assembly is an integral multiple of $\hbar/2$. This situation is included in Van Vleck's (1945) rule of scalar multiplication of spin vectors, in which the parallel/antiparallel spin arrangement, considered above and in section *1.2*, appears as a special case. The exchange energy of two spin vectors S_i and S_j is:

$$E_e = -4S_i \cdot S_j A = -4 S_i S_j A \cos\theta \qquad (1.39)$$

where θ is the angle between spin vectors and A is the exchange integral for half-unit spins ($\hbar/2$) by eq.1.37 and 1.38. The important applications of this more general form are to spin waves and to ferromagnetic domain walls (section *3.3*).

1.4 ANTIFERROMAGNETISM AND FERRIMAGNETISM

Antiferromagnetic materials are paramagnetics in which the spin moments of neighbouring atoms are coupled antiparallel to one another at low temperatures. This is the lowest energy state, because the sign of the exchange integral A (see section *1.3*) is negative. As in ferromagnetism, antiferromagnetic order decreases with increasing temperature and disappears at a critical temperature θ_N, at which thermal energy is equal to the antiferromagnetic exchange energy. This is called the *Néel point* in honour of L. Néel who first postulated antiferromagnetic order and explained the temperature dependence of susceptibility in antiferromagnetic materials. The observation of the Néel points in antiferromagnetics is more subtle than for Curie points in ferromagnetics because θ_N is not characterized by the disappearance of spontaneous magnetization. Two antiferromagnetically coupled sets of atoms (A and B) have a net magnetic moment $I_S = I_A + I_B = (|I_A| - |I_B|)$ but in true antiferromagnetics $I_A = I_B$ and $I_S = 0$. In ferrimagnetics, discussed later in this section, $|I_A| \neq |I_B|$ and there is a net spontaneous magnetization at temperatures below the Néel point. Direct proof of antiferromagnetic order below θ_N is given by neutron diffraction (Shull et al., 1951).

Antiferromagnetics are characterized by their distinctive susceptibility-temperature curves. We follow here Néel's (1948) original theory which is based on the concept of a molecular field. Consider a body-centred cubic lattice (Fig.1.5) in which one set of atoms

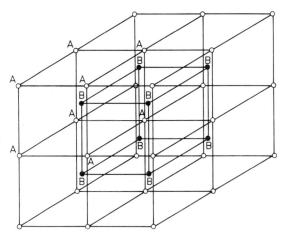

Fig.1.5. Atoms in a body-centred cubic lattice, showing the division into two sub-lattices, *A* and *B*, each of which is itself a simple cubic lattice.

(A), having spins "up" and magnetization I_A, occupies the cube corner sites and another set (B), with spins "down" and magnetization I_B occupies the cube centers. For antiferromagnetism to occur it is necessary that the strongest spin interaction is the negative (antiferromagnetic) one, A – B, between nearest neighbouring A and B sites (or sub-lattices), but there may also be weak intra-sub-lattice (A–A and B–B) interactions between next nearest neighbours. Using the molecular field approximation (as in section *1.2*), the net fields on A and B site atoms are:

$$F_A = -\lambda_{AA} I_A - \lambda_{AB} I_B$$
$$F_B = -\lambda_{BA} I_A - \lambda_{BB} I_B \qquad (1.40)$$

where λ_{ij} denote the molecular field coefficients for the respective interactions. The convention is that the first subscript refers to the atom which sees the field and the second to the source of the interaction field. If A and B sites are occupied by similar atoms, $\lambda_{AA} = \lambda_{BB} = \lambda_i$ and $\lambda_{AB} = \lambda_{BA} = \lambda$. In the presence of an applied magnetic field H, eq. 1.40 become:

$$H_A = H - \lambda_i I_A - \lambda I_B$$
$$H_B = H - \lambda I_A - \lambda_i I_B \qquad (1.41)$$

It is convenient convention to consider the dominant molecular field coefficient λ as a positive quantity, so that to describe antiferromagnetism, it is preceded by a negative sign, whereas λ_i can be positive or negative, because $\lambda_i \ll \lambda$ and its effect on the total alignment is generally small. At thermal equilibrium the magnetizations of the sub-lattices, I_A and I_B are represented by equations of the form 1.18, in which H is either H_A or H_B by (1.41):

$$I_A = \mu B_J \left(\frac{\mu H_A}{kT} \right)$$
$$I_B = \mu B_J \left(\frac{\mu H_B}{kT} \right) \qquad (1.42)$$

At high temperatures (higher that is than the Néel point θ_N which is defined below, eq. 1.50):

$$B_J = \left(\frac{J+1}{3J} \right) \frac{\mu H_{A,B}}{kT} \qquad (1.43)$$

Therefore:

$$I_A = \left(\frac{J+1}{3J} \right) \frac{\mu^2 H_A}{kT} \qquad (1.44)$$

and:

$$I_B = \left(\frac{J+1}{3J} \right) \frac{\mu^2 H_B}{kT} \qquad (1.45)$$

Substituting for H_A and H_B from (1.41):

$$I_A = \frac{\mu^2}{kT}\left(\frac{J+1}{3J}\right)(H - \lambda_i I_A - \lambda I_B)$$
$$I_B = \frac{\mu^2}{kT}\left(\frac{J+1}{3J}\right)(H - \lambda I_A - \lambda_i I_B) \quad (1.46)$$

and the net moment is given by:

$$I = I_A + I_B = \frac{\mu^2}{kT}\left(\frac{J+1}{3J}\right)[2H - (\lambda_i + \lambda)I] \quad (1.47)$$

Then the susceptibility is given by:

$$\chi = \frac{I}{H} = \frac{C}{T + \theta} \quad (1.48)$$

where:

$$C = \frac{J+1}{3J} \cdot \frac{2\mu^2}{k} \quad (1.49)$$

and:

$$\theta = (\lambda + \lambda_i)C/2 \quad (1.50)$$

Eq.1.48 gives the antiferromagnetic susceptibility at high temperatures ($T > \theta_N$) as illustrated in Fig.1.6. Note that λ has been defined as a positive constant, so that θ is a positive quantity. Comparing Fig.1.2 and 1.6 we can see that the sign of the intercept on the temperature axis of a plot of $1/\chi$ vs T gives the sign of the exchange interaction which is responsible for the departure from Curie's law (eq.1.20).

The Néel point, θ_N, is determined by the condition that I_A and I_B, as given by (1.42), both approach zero with H = 0, as for the ferromagnetism, considered in section *1.2*. This leads to the relationship:

$$\theta_N = \frac{C}{2}(\lambda - \lambda_i) \quad (1.51)$$

If $\lambda_i = 0$, θ_N is equal to the Curie-Weiss temperature by eq.1.48–1.50 and the difference between θ_N and θ indicates the magnitude of intra-lattice (AA and BB) interactions. If $\lambda_i \gg \lambda$, the two-sub-lattice model is inadequate and a four-sub-lattice model is required.

At temperatures below the Néel point the susceptibility of an antiferromagnetic material is dependent upon the direction of the field with respect to the axis of spin alignment. In polycrystalline material it is the average susceptibility which is measured, but in a single crystal we distinguish $\chi_\parallel$ and $\chi_\perp$ measured parallel and perpendicular respectively to the spin axis. The application of a perpendicular field deflects the magnetizations I_A and I_B slightly out of perfect antiparallel alignment towards the field direction (Fig.1.7). The angle of deflection ϕ is determined by the balance of the influences of the external field and the molecular field. The molecular field to be considered is only that

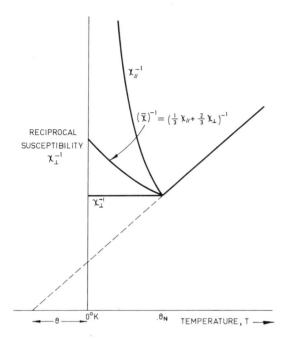

Fig.1.6. Reciprocal susceptibility ($1/\chi$) as a function of temperature (T) for antiferromagnetic materials. Above the Néel point, θ_N, they follow the Curie-Weiss law as for ferromagnetics above their Curie points, except that θ appears as a negative quantity in eq.1.40 when compared with 1.22. The difference is seen by comparing this figure with Fig.1.2. Below the Néel point, antiferromagnetic susceptibility is different parallel and perpendicular to the spin alignment.

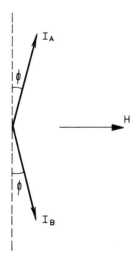

Fig.1.7. Deflection of antiferromagnetic sub-lattice magnetizations in an applied field, H.

between A and B sub-lattices, as represented by the coefficient λ, since the deflection causes no misalignment of spins within each lattice. The total energy, E, of the system is given by:

$$E = -\lambda I_A^2 \cos 2\phi - 2H I_A \sin \phi \qquad (1.52)$$

since $|I_A| = |I_B|$. The equilibrium angle for ϕ is reached when:

$$\frac{dE}{d\phi} = 0 \quad \text{whence} \quad \sin \phi = \frac{H}{2\lambda} I_A \qquad (1.53)$$

Therefore:

$$\chi_\perp = \frac{I_\perp}{H} = \frac{2 I_A \sin \phi}{H} = \frac{1}{\lambda} \qquad (1.54)$$

The other branch of the $1/\chi$ vs T curve in Fig. 1.6 refers to susceptibility parallel to the spin axis at $T < \theta_N$. At $T \to 0°K$, $\chi \to 0$, because $\frac{\mu H_A}{kT}$, $\frac{\mu H_B}{kT}$ in (1.42) tend to infinity. This means that the antiparallel alignment of spins is perfect, independently of H unless $H \approx \lambda M_A$, in which case the external field can over-ride the molecular field. At higher temperatures $\chi_\parallel$ increases smoothly to the value λ^{-1} at $T = \theta_N$. The detailed analysis is given in the original paper by Néel (1948) and also by Smart (1955). In a polycrystalline antiferromagnetic material the susceptibility is close to:

$$\chi = \frac{1}{3}\chi_\parallel + \frac{2}{3}\chi_\perp \qquad (1.55)$$

Although antiferromagnetic materials can be either metallic or insulating, in rock magnetism we are concerned almost entirely with insulating ionic compounds in which the magnetically active metallic ions are separated by magnetically inactive oxygen ions. But although they are not themselves magnetically active, the oxygen ions effect an exchange linkage between the anions, a phenomenon known as superexchange, first suggested by Kramers (1934) and subsequently discussed by numerous authors, especially Anderson (1963) and Goodenough (1963). The type material, in terms of which most fundamental theories of antiferromagnetism have been developed is manganous oxide, MnO, and we here follow Anderson's (1959) descriptive explanation of antiferromagnetic coupling in MnO, appealing to the Pauli exclusion principle and Hund's rule (see section *1.3*).

Consider a total of four electrons on adjacent ions, Mn^{2+}, O^{2-}, Mn^{2+} as in Fig. 1.8. Two electrons are in the completed 2p shell of O^{2-} and one each are in the 3d shells of the Mn^{2+} ions. Pauli's exclusion principle requires that the two O^{2-} electrons in the same orbital state, i.e., with the opposite dumbell-shaped charge distributions represented in Fig.1.8, have opposite spins. These orbitals overlap appreciably with the d-orbitals of the adjacent Mn^{2-} ions and therefore can be regarded as momentarily

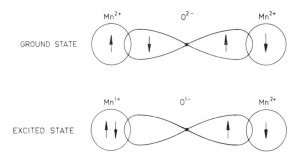

Fig.1.8. Exchange of electrons between manganese and oxygen ions, giving rise to superexchange interaction in MnO.

occupying vacant d orbitals, reducing the ionic charges to Mn^+. But Mn^{2+} is a $3d^5$ ion, i.e., five of the ten 3d states are occupied and by Hund's rule all five 3d electrons have parallel spins, so that a sixth electron must therefore be antiparallel to the spin moment of Mn^{2+} in order to enter a 3d state of Mn. Thus we have two O^{2-} electrons with opposite spins, each coupled to an Mn^{2+} ion in such a way that the Mn^{2+} moments are antiparallel to the two electrons and therefore to one another. Two conditions for effective antiferromagnetic exchange can be seen immediately from this model: (*1*) the anion-cation charge overlap must be appreciable, i.e., the interionic distance must be small, usually of order 2 Å; and (*2*) the angle subtended at the anion by two cations must be close to 180°.

As in the case of ferromagnetism, the antiferromagnetic order-disorder transition at θ_N gives rise to a specific heat anomaly and to changes in lattice parameters. Since antiferromagnetism itself does not cause spontaneous magnetization of a crystal lattice, no remanent magnetism is associated with it, but in rock magnetism we are concerned with two kinds of imperfect antiferromagnetism, that is *canted antiferromagnetism* and *ferrimagnetism*. The best known canted antiferromagnetic is the important mineral hematite (αFe_2O_3), in which equal sub-lattice magnetizations are not quite antiparallel, so that there is a small net magnetization normal to the average spin axis. This is discussed in Chapter 2. Ferrimagnetism occurs when the sub-lattice magnetizations are opposite but unequal. The word was coined by Néel (1948) to describe the magnetism of ferrites, of which the best known is the mineral magnetite, Fe_3O_4. (A valuable reference for the properties of ferrites is Smit and Wijn, 1959). With sub-lattice magnetizations I_A, I_B, the net spontaneous magnetization of the lattices is ($|I_B| - |I_A|$); in magnetite, I_B is due to one ferric ion (Fe^{3+}) with a moment of 5 Bohr magnetons (μ_B) and one ferrous ion (Fe^{2+}), with 4 μ_B, per molecule of Fe_3O_4, and I_A is due to one ferric ion (Fe^{3+}, 5 μ_B) only. The spontaneous magnetization ($|I_B| - |I_A|$) is thus $5 + 4 - 5 = 4 \mu_B$ per molecule at 0°K.

The magnetic properties of ferromagnetics and ferrimagnetics are generally similar; both have spontaneous magnetizations within domains and Curie[1] points at which spontaneous magnetizations vanish. However, they can be distinguished by the temperature dependence of susceptibility above the Curie point; also ferrimagnetics display a wide variety in the curves of spontaneous magnetization vs temperature (I_S vs T) below θ_c, some of which are quite unlike normal ferromagnetics. The temperature-dependence of susceptibility of ferrimagnetics at $T \gg \theta_c$ is similar to that of antiferromagnetics above θ_N:

$$\chi = C/(T + \theta') \tag{1.56}$$

However, as the Curie point is approached, this simple relationship breaks down and appears as the high-temperature asymptote of a more general expression, derived by Néel (1948) from molecular field theory:

$$\frac{1}{\chi} = \frac{T}{C} - \frac{1}{\chi_0} - \frac{\gamma}{T - \theta'} \tag{1.57}$$

where χ_0, γ and θ' are constants which may be expressed in terms of the inter-sub-lattice molecular field coefficient, λ, and the two (now different) intra-sub-lattice coefficients, λ_{AA} and λ_{BB}; the Curie constant C is a sum of the different Curie constants for the two sub-lattices. As in the case of ferromagnetism $\chi \to \infty$ as $T \to \theta_c$.

At $T < \theta_c$, there is a spontaneous magnetization $I_S(T) = (|I_{S_B}(T)| - |I_{S_A}(T)|)$, due to the difference between sublattice magnetizations, which depend upon all three molecular field coefficients λ_{AB}, λ_{AA}, λ_{BB}. For convenience we put:

$$\alpha = \frac{\lambda_{AA}}{\lambda_{AB}} \quad \text{and} \quad \beta = \frac{\lambda_{BB}}{\lambda_{AB}} \tag{1.58}$$

so that the molecular fields F_A and F_B at A and B sub-lattices are:

$$\begin{aligned} F_A &= -\lambda_{AB}(n_B I_B + \alpha n_A I_A) \\ F_B &= -\lambda_{AB}(\beta n_B I_B + n_A I_A) \end{aligned} \tag{1.59}$$

where $n_{A,B}$ denote the fractions of magnetic ions in A and B sites. For three ranges of n_A/n_B, α, β, three different types of $(I_S - T)$ curve occur, as illustrated in Fig. 1.9. Néel referred to these as N, P and L types:

---

[1] In accordance with the usual practice we shall refer to the critical temperature for thermal disorder of ferrimagnetism as the Curie point (θ_c). Some authors call it the ferrimagnetic Néel point, but we believe that confusion is avoided by restricting the term Néel point (θ_N) to pure antiferromagnets.

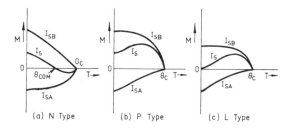

Fig.1.9. Spontaneous magnetization curves for the three types of ferrite recognized by Néel (1948).

N-type: $\quad 1 > \dfrac{n_A}{n_B} > \dfrac{1-\beta}{1-\alpha}$

P-type: $\quad \dfrac{1+\beta}{1+\alpha} > \dfrac{n_A}{n_B} > 1$ (1.60)

L-type: $\quad \dfrac{n_A}{n_B} = 1$

assuming $I_{S_B} > I_{S_A}$ at $0°K$.

N-type ferrimagnetism occurs when β is large and positive, that is λ_{BB} has a negative sign because of antiferromagnetic, intra-sub-lattice superexchange between B site ions, and α is small, so that the molecular field seen at a B site (F_B) is small while that at an A site (F_A) is large. The sub-lattice magnetization I_{S_B} then has a stronger temperature dependence than I_{S_A}. This can lead to a reversal of spontaneous magnetization at a *compensation point* (θ_{com}) as shown in Fig.1.9 (a). At $T > \theta_{com}, I_{S_A} > I_{S_B}$, but at $T < \theta_{com}, I_{S_B} > I_{S_A}$. P-type ferrimagnetism occurs when β is large but negative, that is B–B exchange is positive, and α is small, so that $F_B > F_A$. Then I_{S_B} has a weaker temperature dependence than I_{S_A}, as shown in Fig.1.9 (b). If the ratio $n_A/n_B = 1$, the result can be an L-type $I_S(T)$ curve as shown in Fig.1.9 (c). Although such behaviour can be produced in laboratory specimens of ferrite by careful preparation it is unlikely to occur naturally.

Magnetite and its common impure forms, the titanomagnetites, have spinel crystal structures, in which each A site ion is coordinated tetrahedrally with four oxygen ions, while each B site ion is coordinated octahedrally with six oxygen ions. The various cation–anion distances and the angles subtended at the anions by the cations give two kinds of AB, two kinds of BB and one kind of AA interaction. In magnetite and titanomagnetites there is no 180° AB interaction; the relationship between magnitudes of the interactions is $AB_{126°} \gg BB_{90°} > AA_{79°}$. The other possible interactions, $AB_{154°}$ and $BB_{125°}$ have negligible strengths. The sign of all of the important interactions is negative (i.e., coupling is antiferromagnetic). The signs of the BB and AA interactions are relevant to the apparent P-type magnetization curves observed by Schult (1968) in natural titanomagnetites (see Chapter 2).

1.5 THE ORIGIN OF ANISOTROPY AND MAGNETOSTRICTION

The elementary theory of exchange interaction (section *1.3*), or of superexchange (section *1.4*) implies that it is isotropic, that is, the energy of the parallel or antiparallel spin coupling between two atoms does not depend upon the spin orientation with respect to the line of atomic centres. This is not quite correct: every ferro-, ferri- and antiferromagnetic, has certain crystallographic easy directions along which the spin moments of the ions prefer to lie; they avoid the hard directions. This phenomenon is called magnetocrystalline anisotropy. It is a function of the crystallographic symmetry of a magnetic material. Under an external influence, such as the application of a field, spins may be deflected out of the easy direction but the field must work against the anisotropic constraint, producing anisotropy energy. In Chapter 3 the anisotropy of a titanomagnetite is represented by empirical constants, K_1, K_2, so that the anisotropy energy is quantitatively related to the spin orientation with respect to the crystal lattice. The constants are characteristic of each material and are strongly dependent upon temperature. Here we consider the physical mechanism which gives rise to anisotropy and also to magnetostriction, which is a consequence of anisotropy.

The interaction of the spin moment of an electron with its orbital moment couples the spin moment to the crystal lattice. The spin-orbit coupling has a relativistic explanation (Smit and Wijn, 1959), as does the spin moment itself, but we can envisage a simple magnetic coupling in terms of the field at the electron due to the motion of a charged nucleus relative to the orbiting electron (i.e., relative to the electron the nucleus is orbiting). That the orbital moment is in turn coupled to the lattice can be regarded as part of the process of chemical bonding of the lattice, although the 3d electrons responsible for magnetic properties have only a secondary role in the bonding; it is the 4s electrons which determine the valencies of elements in the first transition series. The energy E_c of the spin-orbit coupling can be expressed in terms of the scalar product of orbital and spin angular momenta, L and S, by an empirical coupling constant:

$$E_c = \alpha' L \cdot S \tag{1.61}$$

Conventionally α' is positive when antiparallel alignment of L and S is favoured. This occurs for electrons in d shells which are less than half filled. In the case of Fe^{2+}, which is a $3d^6$ ion, α is negative, i.e., L and S are coupled parallel. For Fe^{3+}, which is a $3d^5$ ion having a d shell exactly half-filled, $\alpha = 0$, i.e., no anisotropy results from the presence of Fe^{3+} ions.

Van Vleck (1937) considered the anisotropy energy e_k of interaction between two ions to be expanded as a series of even order Legendre polynomials in $\cos \phi$, ϕ being the angle between a principal axis in the lattice and the spin axis:

$$e_k = a + b\left(\cos^2 \phi - \frac{1}{3}\right) + c\left(\cos^4 \phi - \frac{6}{7}\cos^2 \phi + \frac{3}{35}\right) + \ldots \tag{1.62}$$

This is termed the pseudo-dipolar model because if the orbital moments are completely "quenched", i.e., perfectly coupled to the lattice, so that magnetization is entirely a spin effect, eq.1.62 reduces to the first two terms of which the second has the form of the interaction between dipoles; for magnetic dipole interaction:

$$b = -\frac{3m^2}{r^3} \tag{1.63}$$

r being the interionic distance and m the ionic spin moment. Integration of the interaction energies e_k over the whole lattice gives the anisotropy energy E_k which is an important parameter in domain theory (Chapter 3).

Since the anisotropy energy of a material depends upon the angles of the spin moments with respect to the crystallographic axes, and spin waves are departures from perfect spin alignment, it follows that the increase in spin misalignment with increasing temperature causes a reduction in the anisotropy constants, K, which can be represented in terms of the spin misalignment, and consequently directly in terms of spontaneous magnetization, I_S. The result, from spin wave theory is:

$$\frac{K(T)}{K(0°K)} = \left|\frac{I_S(T)}{I_S(0°K)}\right|^n \tag{1.64}$$

where $n \approx 10$ (Callen and Callen, 1963). Values measured for various metals and ionic compounds range between 3 and 10 and for titanomagnetites $n \approx 8.5$ (Fletcher and Banerjee, 1969).

Magnetostriction is a consequence of the strain dependence of magnetocrystalline anisotropy. If the lattice of a magnetic material is deformed, so that the ions are farther apart in a particular crystallographic direction, then some of the interactions giving rise to its anisotropy are diminished and others are increased; if the sign of the resulting energy change is negative then the lattice expansion originally imposed is energetically favoured by the magnetic interactions so that the lattice deforms spontaneously. This is the phenomenon of magnetostriction. Magnetostriction is said to be positive if a material is spontaneously extended in the direction of its magnetization, in which case it is also contracted in the transverse directions. Negative magnetostriction refers to a contraction in the direction of magnetization. Within the domains of spontaneous magnetization in a ferro- or ferrimagnetic the lattice is spontaneously deformed and the deformation becomes apparent in the whole body when it is magnetized by aligning the domains to a common direction.

When the lattice within a domain is constrained not to accomodate its own magnetostriction, for example by neighbouring domains with different orientations in the same crystal, then a magnetostrictive strain energy appears and this is one of the parameters of domain theory (section *3.3*). Such magnetostrictive strain energy is really an additional contribution to the anisotropy energy. A material with positive magnetostriction, such as the titanomagnetites, becomes harder to magnetize in the direction of a compression. This results in the piezomagnetic effect, the converse of magnetostriction, whereby the magnetization of a material, such as a rock, is modified by the application of a stress. This effect is the subject of Chapter 11.

Chapter 2

MAGNETIC MINERALS

2.1 FERRIMAGNETISM OF MAGNETITE

As mentioned in section *1.4* the spontaneous magnetization I_S of magnetite is the net difference between the antiparallel magnetizations I_A and I_B due to iron atoms in the tetrahedrally coordinated A-sites and octahedrally coordinated B-sites of the magnetite lattice. I_S has been measured down to $4.2°K$ in strong applied fields (Pauthenet, 1950). Extrapolated to $0°K$ it corresponds to 4.2 μ_B (Bohr magnetons) per molecule, quite close to 4.0 μ_B, as expected from the Néel model. The difference is probably due to a small orbital contribution to the moment. If the A- and B-site iron atoms were, instead, ferromagnetically coupled, I_S at $0°K$ would have corresponded to $(9+5)\,\mu_B = 14\,\mu_B$. At room temperature ($300°K$), the spontaneous magnetization corresponds to 90 e.m.u./g. Using neutron diffraction the magnetic alignments of the two sublattices can be observed separately. In this way Riste and Tenzer (1961) obtained the temperature-dependences of M_A and M_B for magnetite and compared $(M_B - M_A)$ with magnetic measurements of I_S vs T by Pauthenet (1950). The excellent agreement confirmed the Néel model of ferrimagnetism in magnetite. Another method of determining M_A and M_B separately uses the Mössbauer effect. A Mössbauer absorption spectrum yields the hyperfine magnetic fields (H_{fA} and H_{fB}) at the iron nuclei on A- and B-sites. These fields are proportional to the numbers of outer-shell unpaired electrons, which are also responsible for the atomic magnetic moments (i.e., 9 μ_B per molecule for the B-site iron atoms and 5 μ_B per molecule for the A-site iron atoms at low temperature). Multiplying H_{fA} and H_{fB} by known coefficients, values of M_A and M_B are found at any temperature. Data for magnetite obtained in this way by Van der Woude et al. (1968), shown in Fig.2.1, give independent confirmation of the Néel model. M_B is seen to decrease with temperature more rapidly than M_A. This is consistent with a model in which the strongest superexchange coupling is the negative (i.e., antiparallel) A–B interaction and the smaller B–B interaction between B-site iron atoms is also negative. The negative B–B interaction hastens thermal disorder. We shall use this fact to explain the temperature-dependence of I_S for titanomagnetites in the following section.

Magnetite undergoes a cubic-to-orthorhombic crystallographic transition on cooling through $118°K$. The change is accompanied by onset of electronic order which has important effects on the magnetic and electrical properties. Above the transition temperature, Fe^{2+} and Fe^{3+} ions are randomly distributed on the B-sites which makes electron-hopping between B-site iron ions energetically easy. As an electron hops from a Fe^{2+} ($3d^6$) ion to a Fe^{3+} ($3d^5$) ion the valencies are locally interchanged but, if they are both B-sites, thermodynamic equivalence is maintained. Verwey and Haayman (1941) postu-

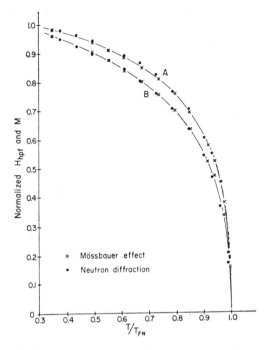

Fig.2.1. Spontaneous magnetization vs temperature for each of the sub-lattices of magnetite, as determined from Mössbauer effect measurements by Van der Woude et al. (1968), and neutron diffraction by Riste and Tenzer (1961).
(Figure reproduced by permission, from Van der Woude et al., 1968.)

lated the electron-hopping mechanism in magnetite to explain its low electrical resistivity ($7 \cdot 10^{-3}$ ohm cm at $300°K$). Below $118°K$, Fe^{2+} and Fe^{3+} ions become ordered in regular arrays along the $\langle 110 \rangle$ axes in the (100) planes as illustrated in Fig.2.2. This causes a sharp increase in the activation energy of electrical conduction from 0.06 eV in the disordered state to 0.5 eV in the electronically ordered orthorhombic state. The cubic $\langle 110 \rangle$ axes become the orthorhombic a and b axes, while the cubic [100] axis [normal to the (100) plane containing the said $\langle 110 \rangle$ axes] becomes the new c-axis.

The crystallographic transition is accompanied by a magnetic isotropic point (i.e., magnetocrystalline anisotropy constant, $K_1 = 0$). Below $118°K$, the B-site Fe^{2+} ions are highly anisotropic due to the partly quenched orbital angular momenta (section 1.5) and are responsible for a large positive K_1, with the easy magnetic axis along the orthorhombic c-axis. Above $118°K$, electron-hopping between Fe^{2+} and Fe^{3+} ions means that there are no true Fe^{2+} ions present and no large anisotropy either. Instead, the magnitude and sign of K_1 are determined by the algebraic addition of K_{1A} and K_{1B} due respectively to the A-site Fe^{3+} ions and the B-site Fe^{3+}-like ions (Wolf, 1957). Wolf has shown on theoretical grounds that K_{1A} of Fe^{3+} ions has a positive sign while K_{1B} for Fe^{3+} is negative.

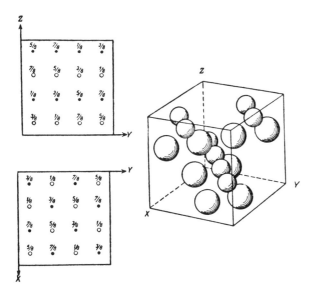

Fig.2.2 Ordered arrangement of the octahedrally coordinated ions in a unit cell of magnetite below 118°K. Large spheres and open circles represent Fe^{2+} and small spheres (and dots) represent Fe^{3+}. Oxygen ions and the tetrahedrally coordinated Fe^{3+} ions are omitted.
(Figure reproduced by permission, from Verwey et al. 1947.)

The net sign of K_1 for magnetite above 118°K is negative, with [111] as the easy magnetic axis. Thus the crystallographic transition at 118°K is a magnetic isotropic point at which $K_1 = 0$ as it changes sign.

The occurrence of magnetic isotropy at 118°K is essential to the phenomena of *magnetic memory* and *low-temperature demagnetization of T.R.M.* in magnetite. Magnetic memory is observed when a rock recovers a fraction of its room temperature value of remanent magnetization after it had lost its remanence by cooling below 118°K and then been warmed through the transition temperature in zero field. It is as if the particles had a memory of the room-temperature remanence in spite of passing through the isotropic point at which there is no directional preference for the magnetization of magnetite. Ozima et al. (1964) reported that stable components of remanence in magnetite survived low-temperature treatment while unstable components were removed, making this a very good method of "magnetic cleaning". Experiments by Merrill (1970) and L. G. Parry (personal communication, 1971) on the nature of thermoremanences of grains in the pseudo-single domain size range (see section *3.6* and *7.4*) showed that true multidomain thermoremanence, which is a bulk property of magnetite grains, is destroyed by cooling through 118°K, but that single-domain and pseudo-single-domain (p.s.d.) remanences are largely recovered on re-warming. This provides a method of examining the nature of p.s.d. moments. In section *3.6* they are attributed to domain effects in the surfaces of small multidomain grains. If this is correct, then the memory effect implies that surface irregularities (and possibly stresses) cause localized pinning of the p.s.d. moments as grains are cooled through the isotropic point. These observations appear not to be consistent with a supposition (Kobayashi and Fuller, 1968) that localized stresses *within the body of a grain* cause the domain pinning which is responsible for the memory effect in magnetite.

2.2 PROPERTIES OF TITANOMAGNETITES

The titanomagnetites ($Fe_{3-x}Ti_xO_4$) are the most common magnetic minerals in basalts. In geological parlance they are frequently referred to simply as magnetites. The substitution of Ti ions for Fe ions in the magnetite lattice, with conservation of total ionic charge, means that (Ti^{4+} and Fe^{2+}) replace 2 Fe^{3+}, i.e., substitution of Ti^{4+} for Fe^{3+} requires also the conversion of a second Fe^{3+} ion to Fe^{2+}. Spontaneous magnetization, I_s, Curie point, θ_c, anisotropy constants, K_1, K_2, coercive force, H_c, magnetostriction constants, λ_1, λ_2, and electrical conductivity all vary in a regular manner with the composition parameter x from $x = 0$ (magnetite) to $x = 1$ (ulvöspinel). (Table 3.1 lists values of some of these parameters at laboratory temperature.) O'Reilly and Banerjee (1965) measured saturation magnetization of titanomagnetites at 77°K in fields up to 30 kilo-oersteds and also the activation energy of electrical conduction, E_ρ, in the ionically disordered (high-temperature) state. Results are plotted in Fig. 2.3, with the spon-

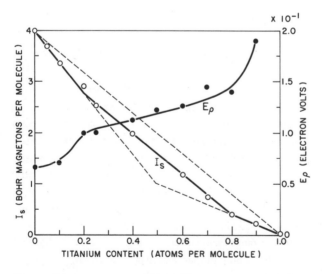

Fig.2.3. Spontaneous magnetization at 0°K and activation energy for electron jumps between Fe^{3+} and Fe^{2+} ions in magnetite.
(From data by O'Reilly and Banerjee, 1965.)

taneous magnetization, I_S, extrapolated to $0°K$. These data indicate that for $0.2 < x < 0.8$ the extra Fe^{2+} ions resulting from titanium substitution enter A-sites preferentially, leaving the Fe^{2+} content of the B-sites unchanged. The cation distribution thus deduced is:

$$0 < x < 0.2 \; : \quad Fe^{3+} \, [Fe^{3+}_{1-2x} \, Fe^{2+}_{1+x} \, Ti^{4+}_x]$$
$$0.2 < x < 0.8: \quad Fe^{3+}_{1.2-x} \, Fe^{2+}_{x-0.2} \, [Fe_{0.8-x} \, Fe^{2+}_{1.2} \, Ti^{4+}_x] \quad\quad (2.1)$$
$$0.8 < x < 1 \; : \quad Fe^{3+}_{2-2x} \, Fe^{2+}_{2x-1} \, [Fe^{2+}_{2-x} \, Ti^{4+}_x]$$

where ions within the square brackets are the octahedrally coordinated B-site ions and the others are the tetrahedrally coordinated A-site ions. Stephenson (1969) suggested that the second of the above distributions represents thermal disorder. As in magnetite, electrical conduction is primarily due to electron hopping between Fe^{2+} and Fe^{3+} ions in B-sites, so that at $x = 0.8$ (above which there are no B-site Fe^{3+} ions) there is a sharp increase in E_ρ.

Banerjee et al. (1967) measured coercive force, H_c as a function of x at $77°K$ and Syono (1964) measured the magnetocrystalline anisotropy constants K_1, K_2, and magnetostriction constants, λ_{100}, λ_{111} between $77°K$ and $300°K$ for a few selected titanomagnetite single crystals up to $x = 0.68$. All of these parameters increase both with increasing x and decreasing temperature. Laboratory temperature values are listed in Table 3.1. Banerjee et al. (1967) explained these observations as due directly to the increase in Fe^{2+} ions per molecule associated with increasing Ti^{4+} concentration. For $x = 1.0$ (ulvöspinel), H_c can be as high as 8,000 Oe for a polycrystalline sample at $77°K$ and unless the applied field is at least two or three times greater than this, it is possible to obtain anomalous "humps" in I_S vs T curves. At $300°K$ the contribution of Fe^{2+} ions to K_1 (and also H_c) may be small but it increases so sharply on cooling that laboratory fields are inadequate to saturate the magnetization completely. This causes an apparent reduction in I_S as the temperature (T) decreases and a "hump" appears in the I_S vs T curve. Such "humps" for titanomagnetites with $x > 0.6$ have been erroneously attributed to an intrinsic P-type behaviour (section 1.4) in titanomagnetites (e.g., Akimoto, 1962). If P-type behaviour is hypothesized as a consequence of limited Ti^{4+} substitution in magnetite, then it follows from Néel's theory that further Ti^{4+} will cause a N-type I_S vs T curve, i.e., a spontaneous self-reversal of remanent magnetization with decreasing temperature. Our conclusion that the apparent P-type behaviour in pure titanomagnetites is an artefact and that there is no possibility of spontaneous self-reversal of magnetization in stoichiometric titanomagnetites is therefore important. It is strengthened by the fact that the intra-sub-lattice B—B interaction in magnetite is negative (section 2.1) and, therefore, P-type I_S vs T behaviour is not to be expected (Blasse, 1964).

The additional Fe^{2+} ions introduced into the titanomagnetite lattice by Ti^{4+} substitution have a strong effect on the low-temperature magnetic transitions (Syono, 1965). With small impurities ($x < 0.3$), the transition temperature ($118°K$ at $x = 0$) is lowered, while for $x > 0.5$, the transition is between $200°$ and $300°K$.

The electrical conductivity transition in magnetite is completely suppressed when $x > 0.1$, because the substituted Ti^{4+} ions maintain ionic disorder of B-site ions. When cooled to temperatures low enough to produce an ordered arrangement of Fe^{2+} and Fe^{3+} ions, there is a residual disorder due to the randomly located Ti^{4+} ions sufficient to suppress the transition in electrical conductivity.

Finally, mention should be made of the unexpected, weak ferromagnetism (i.e., a small but finite value of I_S) in the $x = 1$ end member of the series, ulvöspinel (Fe_2TiO_4). The cation distribution is Fe^{2+} [Fe^{2+} Ti^{4+}] where, as before, the brackets denote octahedral B-sites. The Fe^{2+} ($3d^6$ ion) has a magnetic moment of 4 μ_B (Bohr magnetons) and since the atoms are coupled antiferromagnetically, the net moment, I_S, should be $4 - 4 = 0$ μ_B. In addition to a finite value of I_S, Ishikawa (1967) observed a "hump" in the I_S vs T curve of a single crystal ulvöspinel, the maximum I_S being 0.36 μ_B per molecule. Readman et al. (1967) measured I_S and susceptibility (χ) vs T, as well as Mössbauer absorption in Fe_2TiO_4 and concluded that the orbital angular momentum in the B-site Fe^{2+} ion is only partially quenched while that in the A-site Fe^{2+} ion is fully quenched. The result is that while the A-site Fe^{2+} ion in ulvöspinel has a magnetic moment of 4 μ_B, the magnetic moment in the B-site Fe^{2+} ion (due to spin and some orbital momentum) is 4.5 μ_B. In very strong fields and close to 0°K the value of I_S for ulvöspinel should, therefore, be 0.5 μ_B per molecule.

2.3 MAGHEMITE

Maghemite is so called because although its chemical composition is identical to that of hematite (ferric oxide or α-Fe_2O_3), unlike hematite, it is strongly magnetic (like magnetite), and also its crystal structure is similar to that of magnetite. In order to distinguish it from hematite, the chemical formula is given as γ-Fe_2O_3 to denote its cubic spinel structure (with a tetragonal super-lattice).

The most important property of maghemite is that it is thermodynamically metastable and converts to the more stable rhombohedral hematite structure on heating to temperatures above 350°C. The transition depends upon grain size, water content, foreign ion impurities etc. and is considered further in section 5.5. Here we are concerned with the formation and natural occurrence of maghemite and its magnetic properties.

Maghemite is formed by the low-temperature (150°C—250°C) oxidation of magnetite. Because of the instability, single crystals larger than 1 micron (= 10^{-4} cm) can rarely be synthesized or found in nature. The oxidation is achieved by a topotactic process, so called because the anionic (oxygen) structure is left unchanged; the Fe^{2+} ions diffuse to the surface of the grain, oxidize and form Fe^{3+} ions, leaving lattice vacancies. If the diffusion can proceed unhindered, one microcrystal of magnetite will be completely transformed to maghemite without an additional new crystal of maghemite being formed. Extensive work at the research laboratory of Montecatini-Edison (see, for example, Colombo et al., 1964; Gazzarrini and Lanzavecchia, 1969) has shown, however, that if a minute amount of hematite impurity is present in the initial magnetite grain, two competitive processes occur. In addition to the slow conversion from magnetite to maghemite there is an autocatalytic growth of hematite which becomes predominant in a mixed crystal. Stacking faults in the initial magnetite are equivalent to hematite nucleii, so that in the presence of stacking faults, oxidation produces a mixture of magnetite, maghemite and hematite. However, when some lattice sites are initially vacant, the formation of maghemite is accelerated; also water incorporated as protons in the lattice has the same effect because in the maintenance of charge balance, they encourage the presence of lattice vacancies.

For each Fe^{2+} ion diffusing out of a magnetite lattice, two more are converted to Fe^{3+} to maintain constant total cationic charge within the structure. Thus, $Fe_2^{3+} Fe^{2+} O_4$ becomes $Fe_{8/3}^{3+} \square_{1/3} O_4$, i.e., one third of the original Fe^{2+} sites become vacancies, represented by $\square$. This way of representing the chemical formula also indicates that the crystal structure is similar to the cubic spinel structure of magnetite ($Fe_3 O_4$). The structure is not exactly that of a spinel because the vacancies are ordered along a particular ⟨100⟩ axis, the repeat-distance being three times the cubic cell-edge. The crystal structure is, therefore, tetragonal due to the vacancy superlattice. Although the cubic cell-edges of magnetite (8.39 Å) and maghemite (8.34 Å) are similar, it is possible to distinguish in the X-ray diffraction pattern of maghemite, tetragonal lines which are forbidden in magnetite.

The spontaneous magnetization, I_S, of maghemite is 85 e.m.u./g at 300°K, quite close to that of magnetite. Low-temperature (4.2°K) saturation magnetization and Mössbauer effect measurements show that the lattice vacancies occur in B-sites.

Maghemite has been seen in some red sandstones, oceanic red clays and oxidized basalts, where it has been formed at low temperatures, i.e., by authigenesis or weathering. Maghemite-bearing rocks therefore have secondary components of natural remanent magnetization (N.R.M.) of chemical origin. It is possible to distinguish the presence of maghemite by heating rocks to temperatures above 350°C. Metastable maghemite converts to hematite and saturation magnetization decreases to 0.4 e.m.u./g. Because of this, it has not been possible to determine the true Curie point, θ_c, of maghemite. Various extrapolation methods have been used to determine θ_c from I_S measurements made *below* the crystallographic γ-to-α transition. Brown and Johnson (1962) fitted the low-temperature I_S data to a Brillouin curve which gave θ_c as 750°C. Banerjee and Bartholin (1970) obtained nearly the same value by extrapolating low-temperature I_S data according to the Landau and Lifshitz theory of second-order transitions. The magnetocrystalline anisotropy constant, K_1, has been determined from polycrystalline samples using ferromagnetic resonance, and was found to be $-1.1 \cdot 10^5$ ergs/cm^3, very close to the value for magnetite. The similarity of I_S and K_1 for maghemite and magnetite ensures that theories of magnetic properties of magnetite are generally applicable to maghemite also (except, of course, that maghemite cannot have T.R.M. unless it is doped with impurities which inhibit the transition to hematite).

2.4 OXIDIZED TITANOMAGNETITES

Oxidized titanomagnetite compositions are denoted by the shaded area in Fig.2.4. As their name signifies, these are non-stoichiometric compounds with cation-vacancies, due either to the low-temperature ($< 600°C$) oxidation of some of the Fe^{2+} ions in

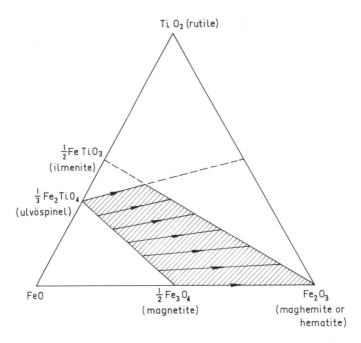

Fig.2.4. Ternary composition diagram for $FeO-Fe_2O_3-TiO_2$ with shaded area showing the range of compositions of oxidized titanomagnetites and arrows indicating the direction of the compositional change during oxidation.

what were originally stoichiometric titanomagnetites, or because the cation-deficient compositions were stabilized at moderate-to-high temperatures ($> 600°C$) during cooling in the presence of various stabilizing magmatic gases. They are of single-phase cubic crystal structure but metastable like maghemite. On heating in air to temperatures above 200°C, an oxidized titanomagnetite decomposes into a two-phase structure, a titanium-poor titanomagnetite with exsolved titanium-rich titanohematite or ilmenite lamellae. The decomposition is a thermally activated rate process with an activation energy of about 0.03 eV as determined by Ozima and Ozima (1972). Similar values of activation energy for electron hopping in titanomagnetites were reported by Banerjee et al. (1967), which suggests that electron hopping activates the thermal decomposition of oxidized titanomagnetites. The chemical formula for the single-phase oxidized titanomagnetites is readily deduced, by starting with the unoxidized titanomagnetite: $Fe^{3+}_{2-2x}Fe^{2+}_{1+x}Ti^{4+}_xO^{2-}_4$

and replacing y Fe^{2+} ions per molecule with $\frac{2}{3} y$ Fe^{3+} ions, leaving $y/3$ lattice vacancies:

$$Fe^{3+}_{2-2x+(2/3)y} Fe^{2+}_{1+x-y} Ti^{4+}_{x} \square_{y/3} O^{2-}_{4}$$

Oxidized titanomagnetites have sometimes been referred to as *titanomaghemites* or the γ-series of iron-titanium oxides. However, by analogy with the so-called α-series (titanohematites) and the β-series (titanomagnetites), the γ-series (titanomaghemites) are strictly oxides in which (Ti^{4+} + Fe^{2+}) replaces $2Fe^{3+}$ in maghemite. Representing maghemite by the formula which implies four oxygen ions per molecule and indicates the proportion of lattice vacancies, $Fe^{3+}_{8/3} \square_{1/3} O^{2-}_{4}$, and introducing xTi^{4+} ions per molecule we obtain the general formula for titanomaghemite, $Fe^{3+}_{8/3-2x} Fe^{2+}_{x} Ti^{4+}_{x} \square_{1/3} O^{2-}_{4}$. This coincides with the general formula above for oxidized titanomagnetites only if $y = 1$, i.e., oxidation is complete. In Fig.2.4 the titanomaghemites are represented by compositions along the hematite-ilmenite join, the difference being that their crystal structure is cubic. The oxidized titanomagnetite compositions are, however, denoted by any point in the shaded area.

The majority of the magnetic minerals responsible for natural remanence in basalts are oxidized titanomagnetites. Akimoto et al. (1957) made the first attempts to produce these minerals in the laboratory by oxidizing synthetic titanomagnetites in air at temperatures between 400°C and 550°C. They measured the spontaneous magnetizations, I_S, Curie points, θ_c, and lattice parameters, a, of the resulting compounds and plotted the apparent increase in all three parameters with oxidation. However, O'Reilly and Banerjee (1967) drew attention to the fact that this heat-treatment always results in two crystallographic phases and more recently Ozima and Sakamoto (1971) studied the X-ray diffraction data of the original samples of Akimoto et al. and conclusively showed that they are two-phase samples. Thus the I_S, θ_c and a-contours of Akimoto et al. are all incorrect. They have now been superseded by the measurements of O'Reilly and Readman (1971), reproduced in Fig.2.5. The work of Gazzarrini and Lanzavecchia (1969) shows that low-temperature oxidation of magnetite to the cation-deficient maghemite structure is possible only when the starting material initially contains cation vacancies which can be stabilized by incorporated water. Sakamoto et al. (1968) and Ozima and Larson (1970) have therefore tried to repeat the low-temperature oxidation experiments of Akimoto et al., using fine-particle titanomagnetites produced by preliminary ballmilling in the presence of water. The results show that I_S decreases and θ_c increases with oxidation as expected from theoretical models, described below. Unfortunately, however, the ballmilling process results in a very wide range of particle sizes, including superparamagnetic particles with different blocking temperatures. Thus the experimentally derived I_S and θ_c values are inaccurate to the extent that the superparamagnetic component contributes to the temperature dependence of the measured saturation magnetization. In the absence of well-characterized samples we are limited to a discussion of the various theoretical models for cation distribution and the resultant magnetic properties expected in oxidized titanomagnetites.

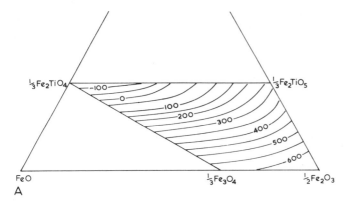

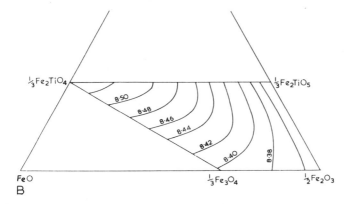

Fig.2.5. A. Curie points (°C). B. unit cell parameter (Å) for oxidized titanomagnetites. (Reproduced by permission from O'Reilly and Readman, 1971.)

Verhoogen (1962) first pointed out that some oxidized titanomagnetites were likely to have a net spontaneous magnetization due to the tetrahedral sub-lattice moment, M_A exceeding the octahedral sub-lattice moment M_B. Thus a stoichiometric titanomagnetite, in which B-site ions dominate the net magnetization, could, after a suitable degree of oxidation, acquire a dominant A-site magnetization, i.e., a reversal in direction of the net moment $|M_B - M_A|$. O'Reilly and Banerjee (1966) criticized the Verhoogen model because of its dependence upon the Néel-Chevallier model for cation distribution in titanomagnetites and the assumption that Fe^{3+} ions always have an overwhelming preference for the A-sites. It is probable that oxidation results first in a decrease in I_S, followed by an increase as exsolution occurs. The Curie point, θ_c, is also expected to increase, so that titanomagnetites with θ_c originally below room temperature will transform to material with θ_c above room temperature. Thus low-temperature oxidation can result in a chemical remanence in basalts. Since the oxidation also causes a decrease in the number of magnetically anisotropic Fe^{2+} ions in the lattice, coercive force, H_c, is expected to be lowered, and hence the new C.R.M. is expected to be relatively unstable. All of these conjectures have yet to be proved, using well-characterized synthetic samples with controlled oxidation.

2.5 WEAK FERROMAGNETISM IN HEMATITE

Hematite (α-Fe_2O_3) and titanohematite ($Fe_{2-x}Ti_xO_3$) are common magnetic minerals in both sedimentary and igneous rocks. Hematite crystallizes in the trigonal system and is often indexed as a rhombohedral or hexagonal mineral; thus the trigonal c-axis is also called the rhombohedral [111] axis or the hexagonal [0001] axis. The iron ions, being all Fe^{3+}, have magnetic moments of 5 μ_B and are all octahedrally coordinated. The magnetic moments of the ions within a specific c-plane are ferromagnetically coupled (i.e., parallel to one another) but anti-ferromagnetic coupling between the planes couples them in parallel pairs, but with alternating polarity of the pairs of planes. The spin alignment is thus antiferromagnetic, but the antiparallelism is imperfect. Above $-10°C$, the spin moments are oriented in the c-plane but instead of being precisely antiparallel, they are slightly canted (Dzyaloshinsky, 1958; Moriya, 1960), resulting in a weak spontaneous magnetization (I_S = 0.4 e.m.u./g) within the c-plane, but normal to the spin-axis. Below $-10°C$, due to a change in the sign of the magnetocrystalline anisotropy, the c-axis becomes the spin axis. The mutual orientation in a unit cell is still the same but the spins are then parallel or anti-parallel to the c-axis. There is then no spin canting so that I_S = 0, i.e., at low-temperatures hematite is a perfect antiferromagnetic. The spin-axis or "spin-flop" transition in hematite is called the Morin transition after F. J. Morin, who investigated it in great detail.

The canting of the antiferromagnetic spin-axis has been directly observed by Nathans et al. (1964) using polarized-neutron diffraction but it is suspected that secondary contributions to the spontaneous magnetization of hematite arise from defects or impurities preferentially occupying one sub-lattice or the other. Slight, but variable spontaneous magnetization, not explained by the canting theory, is observed in the c-plane below the Morin transition and along the c-axis above the transition and is correlated with the introduction of defects (impurities or by neutron bombardment). Néel's (1953) original idea that stacking faults in the hematite lattice may cause small, strongly ferrimagnetic regions may, therefore, be correct, although spin canting accounts for the major part of hematite ferromagnetism.

Above the Morin transition, and below the Néel point, θ_N, the weak ferromagnetism is constrained to lie in the c-plane by what is in effect a very strong uniaxial anisotropy; by comparison, anisotropy in the c-plane is very small and variable. Direct measurements of c-plane anisotropy have been made using the static techniques of torque magnetometry and the Mössbauer effect and the dynamic technique of ferromagnetic resonance (see Chapter 5). In addition to the anisotropy of triaxial (6 θ) symmetry expected in the c-plane of a trigonal crystal, uniaxial (2θ) and cubic (4θ) components are observed. The triaxial anisotropy constant, K_3, is variable from crystal to crystal, and is of order 10^2 ergs/cm^3. This is much too small to explain the high coercive forces found in hematite, of the order of 1000 Oe for 15 μm grains. The additional uniaxial and cubic anisotropies which are unrelated to the crystal symmetry, but are caused by the magnetostrictive energy of oriented defects, account for the magnetic hardness. Hematite has a saturation magnetostriction $\lambda = 8 \cdot 10^{-6}$ and Porath and Raleigh (1967) have shown that twinning in the c-plane can produce anisotropy sufficient to cause high coercivity. This accounts for stable remanence in hematite-bearing rocks.

In magnetite, the net magnetocrystalline anisotropy constant, K_1, is an algebraic sum of contributions K_{1A} and K_{1B} from the two, differently coordinated sub-lattices. Although the iron atoms in hematite are all in octahedrally coordinated (B) sites, there are still two independent contributions to the uniaxial (out-of-plane) anisotropy constant, K. One of these is the magnetic dipolar contribution, K_{MD}, which tends to align the spins parallel to one another in the c-plane. The other is the so-called fine structure contribution, K_{FS}, which is due to the electrostatic fields of oxygen anions, and tends to orient the spins along the c-axis. K_{MD} and K_{FS} have different temperature-dependences; below the Morin transition K_{FS} predominates and the spins are parallel to the c-axis while above the Morin transition, K_{MD} predominates and, therefore, the spins are parallel to the c-planes. Application of pressure or of a magnetic field, the presence of impurities or reduction in grain size, all affect the balance between K_{FS} and K_{MD} and shift the Morin transition temperature (T_M). The pressure-dependence, $dT_M/dP = + 3.6°$/kbar, is such that at about 30 kbar, T_M is raised to room temperature (300°K). On the other hand, impurities like Ti^{4+} and Sn^{4+} depress T_M; 1% Sn^{4+} reduces T_M to $-40°$C.

The Morin transition is a point of magnetic isotropy ($K = 0$) in hematite, as is the 118°K transition in magnetite. And, as in magnetite, the remanent magnetization of hematite shows a memory phenomenon when cooled and then warmed through T_M in a field-free space. As the temperature is lowered through T_M, the magnetization decreases sharply to very low values (1-10% of the room temperature value), but, on warming, a substantial fraction of the room-temperature remanence is recovered. As in the case of magnetite, the explanation appeals to microscopic regions which retain the room temperature anisotropy in spite of cooling, perhaps by virtue of local internal stresses (Kawai et al., 1968; Kobayashi and Fuller, 1968), but the detailed mechanism is not yet clear. For example, the use of magnetic colloids has shown that even below T_M, some colloidal patterns remain around impurity- or defect-sites in a single crystal. When the crystal is warmed, colloidal patterns spread from the nucleation sites to the whole of the crystal.

2.6 TITANOHEMATITES

Titanohematites are represented by the chemical formula $Fe_{2-x}Ti_xO_3$ where x varies between 0 and 1. They crystallize in the trigonal system but for $0 < x < 0.5$ the space group is $R\bar{3}C$ while for $0.5 < x < 1.0$ the space group is $R\bar{3}$. Solid solution is complete only above 700°C so that some compositions can be obtained only by quenching from above 700°C. Slow cooling results in exsolution of hematite-rich ($0 < x < 0.2$) and ilmenite-rich ($0.8 < x < 1.0$) components, known as ilmeno-hematites and hemo-ilmenites respectively, which are quite common in igneous rocks and as igneous detritus in sedimentary rocks. Single-phase titanohematites with $0.2 < x < 0.8$ are rare but have been seen in quenched lavas. The compositions around $x = 0.5$ have unusual magnetic and crystallographic properties, which include reversal of remanence (Chapter 12).

In ilmenite, as in hematite, the predominant spin coupling is antiferromagnetic between c-planes but the individual ion moments are directed parallel and antiparallel to the c-axis. Ilmenite is a perfect antiferromagnet in which the moments of Fe^{2+} ions are exactly mutually cancelling. The variation of spontaneous magnetization with composition for intermediate compounds in the hematite-ilmenite series is shown in Fig.2.6.

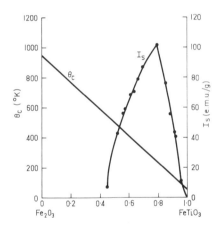

Fig.2.6. Spontaneous magnetizations I_S and Curie points θ_c of the hematite-ilmenite solid solution series, based on data by Bozorth et al. (1957), Ishikawa and Akimoto (1958) and Nagata and Akimoto (1956).

The same diagram also shows the smooth variation of Néel point from 953°K to 50°K. The magnetocrystalline anisotropy constant, K_1' has not been measured for any of these trigonal crystals, but the coercive force, H_c is plotted in Fig.2.7 as a function of Ti^{4+}-substitution for the range with Néel points above 300°K. It has a peak at $x = 0.2$, attributed to magnetostrictive strain due to microscopic exsolution lamellae of $Fe_{1.8}Ti_{0.2}O_3$ in a matrix of $Fe_{1.2}Ti_{0.8}O_3$ (Merrill, 1968). The lattice parameters of these two compounds are different and it has been suggested that the lattice mismatch causes strain and asso-

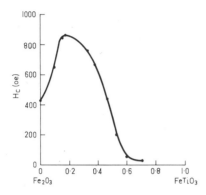

Fig.2.7. Coercive forces of hematite-ilmenite solid solution series, from data by Nagata and Akimoto (1956).

ciated magnetostrictive anisotropy, causing high coercive forces although Shive and Butler (1969) have shown that this is not the complete explanation.

Searle and Morrish (1966) suggested that the additional Fe^{2+} ions accompanying the substituted Ti^{4+} ions in titanohematites may form a third sub-lattice with moments antiparallel to the Fe^{3+} moments so that the canted Fe^{2+} moment progressively cancels the canted Fe^{3+} moment. If correct this has an important implication which appears to have escaped notice. The Fe^{2+} and Fe^{3+} canted moments will be different functions of temperature, so that precise mutual cancellation will occur at a particular compensation temperature θ_{com} as in the case of N type ferrimagnetics (section 1.4). The value of θ_{com} will depend upon the Fe^{2+}/Fe^{3+} ratio, i.e., upon the degree of Ti^{4+} substitution. This provides a possible mechanism for self-reversals (see Chapter 12) and could also account for the very high coercivities (H_c) of titanohematites with $x \approx 0.2$, since $H_c \propto K/I_S \to \infty$ as $I_S \to 0$ for finite K (K being anisotropy energy and I_S is spontaneous magnetization).

Independently of the above postulated mechanism for self-reversal in titanohematites with $x \approx 0.2$, titanohematites with $x \approx 0.5$ are known to undergo self-reversal by a process of ionic ordering (Chapter 12). As noted above, at $x = 0.5$ the titanohematites change their space grouping from $R\overline{3}C$ to $R\overline{3}$. If a crystal of composition $x = 0.5$ is cooled slowly from above its order/disorder temperature, it is possible to obtain either pure $R\overline{3}C$ or $R\overline{3}$ symmetry throughout the whole crystal. However, if the cooling is rapid, only short-range order occurs creating iron-rich and iron-poor regions in the crystal. Such regions occur for titanohematites with $0.4 < x < 0.6$ if they are not allowed to reach equilibrium by previous high temperature annealing. As a result it is possible to have a macroscopically single-phase chemical composition in this range with variable short-range order. The boundaries between these ordered regions are crystallographically coherent, so that magnetic superexchange interactions couple them strongly and cause self-reversal of remanence (section 12.2).

2.7 MINOR MAGNETIC MINERALS

Apart from the iron-titanium oxides there are two mineral types with intrinsic ferromagnetic properties, sulphides, with a range of compositions between troilite, FeS and pyrrhotite Fe_7S_8, and the hydrated oxide, goethite or α ferric oxyhydroxide (αFeOOH). Numerous other minerals contain dissolved iron oxides which may slowly exsolve at low temperatures to produce very fine magnetic grains and, if they exceed the critical size for magnetic stability (Chapter 6), impart a stable remanence to intrinsically non-magnetic minerals. This has been shown to explain the magnetic properties of numerous minerals (Hanuš and Krs, 1965) including felspars (Evans and McElhinny, 1968; Hargraves and Young, 1969), zircon (Lewis and Senftle, 1966) and cassiterite, SnO_2 (Banerjee, 1969; Banerjee et al., 1970). Although in a few such cases the properties may indicate more complex compounding of iron oxides with the host minerals, in general the magnetic properties are due to traces of the well-known iron-titanium oxides described in the earlier sections of this chapter.

Troilite, FeS, is a hexagonal, antiferromagnetic mineral with a Néel point of 320°C. Being perfectly antiferromagnetic it is of little interest in rock magnetism. However, pure troilite is found only in meteorites in which an excess of metallic iron ensures the stoichiometric composition FeS. In ore bodies and in rock formations, it is more common to observe the iron-deficient sulphides, the most iron-deficient being pyrrhotite which may be represented by $Fe_{0.875}\square_{0.125}S$, in which the square denotes a cation vacancy. The vacancies in pyrrhotite form an ordered structure and are located preferentially in alternate iron layers normal to the hexagonal symmetry axis of FeS (Fig.2.8). The ordered arrange-

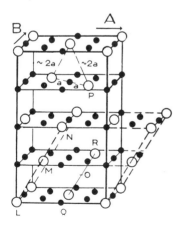

Fig.2.8. Lattice vacancies in pyrrhotite represented by open circles, with solid circles giving the positions of Fe^{2+} and Fe^{3+} ions. In troilite (FeS) the lattice sites are all occupied and all cations are Fe^{2+}.
(Figure reproduced by permission, from Bertaut, 1952.)

ment of the vacancies has two effects: (a) the hexagonal crystallographic symmetry is slightly distorted to form a monoclinic structure; and (b) due to the inequality in numbers of ions on the antiferromagnetically coupled sub-lattices a net ferrimagnetic moment is present. The spontaneous magnetization of pyrrhotite is 13.5 e.m.u./g at 20°C and its Curie point is 320°C (the Néel point of troilite). The magnetocrystalline anisotropy constant in pyrrhotite is negative; the spins are constrained to lie in the c-plane normal to the hexagonal c-axis, as in hematite above the Morin transition.

Although it is fairly easy to understand the magnetic properties of the two end members of the iron sulphide series, troilite and pyrrhotite, it is more difficult to characterize the magnetic properties of the intermediate members, which have different vacancy concentrations. But in nature the intermediate compositions are quite common, as was shown by the work of Everitt (1962b) and Schwartz (1968). For vacancy concentrations less than 0.125 per molecule, a disordered arrangement is obtained if the material is quenched from a high temperature (Schwartz, 1968). A completely random distribution of the vacancies between the anti-ferromagnetically coupled sub-lattices results in a zero net magnetization in the quenched state. When re-heated slowly, however, there is enough thermal energy at about 200°C for the vacancies to order and locate themselves preferentially on one iron sub-lattice, as in pyrrhotite. The thermomagnetic curve of such a quenched sample, therefore, shows a sudden rise in magnetization at about 200°C, followed by a decrease to zero at 320°C, the Curie point of pyrrhotite, at which long-range ferrimagnetic order disappears.

It is usual in nature to find quenched sulphide with a gradation of vacancy concentrations. When annealed at temperatures of the order of 300°C, two processes occur simultaneously. First, the disordered vacancies in the intermediate $Fe_{1-x}\square_x S$ phases approach ordered arrangement, producing a net ferrimagnetic moment in previously antiferromagnetic materials. Second, the different chemical phases are homogenized to produce a single chemical composition with a common vacancy concentration. It is possible for either method to cause self reversal of remanence.

Goethite (α-FeOOH) is an antiferromagnetic mineral which ideally should have perfectly compensated spins and hence no net spontaneous moment. However, it has been reported that both natural and synthetic goethite are capable of acquiring a thermoremanent magnetization (TRM) when cooled through the Néel point. The reason for this lies in oxygen-ion vacancies in goethite which cause the formation of broken links in the continuous antiferromagnetic iron-oxygen-iron chains. Since the vacancies occur at random, some of the broken links can contain an odd number of antiparallel iron atoms and hence such links can carry spontaneous magnetic moments. When such a goethite is cooled through the Néel point the anisotropy is small, and hence the moment-carrying links line up along the applied field direction and result in a TRM. Since goethite occurs in red sandstones and shales, such an acquisition of stable TRM at a moderately low temperature ($\sim 100°C$, the Néel point of goethite) is feasible in these rocks.

Chapter 3

FERROMAGNETIC DOMAIN THEORY

3.1 MAGNETOSTATIC ENERGY

The magnetostatic energy between neighbouring atoms in a ferromagnetic material is smaller than their mutual exchange energy by a factor of 1,000 or more, as noted in section *1.3*. However, the exchange forces are short-range forces acting only between nearest and next-nearest atomic neighbours, whereas magnetostatic forces are effective on a macroscopic scale. They perturb the perfect spin alignment favoured by the exchange forces and produce a domain structure, with regions or domains inside which there is parallel spin alignment, but which are separated by relatively narrow domain walls through which there is a progressive rotation of spins. In the absence of external constraints the spin alignments or magnetizations of the domains are so arranged as to form loops of magnetic flux closure within a magnetic body, such as a grain of magnetite, as in the simple example of Fig.3.2(b). In this state the body is unmagnetized, that is, it has no externally observed magnetic moment, although it may become magnetized by exposure to an external field which upsets the balance of internal magnetostatic energy.

There are two distinct contributions to the magnetostatic energy of a body. One is its potential energy in an external field and the other is the mutual potential energy of its surfaces of magnetic polarity. For convenience we use the term magnetostatic energy and the symbol E_m for the second of these and refer to the first as the external field energy. Since our concern is normally with an external field H which is uniform over the dimensions of any body on which we have to consider magnetostatic forces, the external field energy, E_H, is

$$E_H = -H \cdot M = -HM \cos\theta \tag{3.1}$$

where the total or net magnetic moment of the body, M, makes an angle θ with the field H.

The magnetostatic energy may be written as a double integral over all surfaces of magnetic polarity of the mutual potential energy of two elementary areas of surface. Normally, magnetic bodies do not have internal surfaces with significant net polarities because orientations of domain walls are selfadjusting to bisect the angle between the magnetization directions of adjacent domains, as in Fig.3.1. The double integral then applies only to the external surface. Sub-division of a ferromagnetic material into domains tends to eliminate also magnetic polarity from its external surface thus reducing its magnetostatic energy. A simplified example is given in Fig.3.2.

Analytical expressions for magnetostatic energies of bodies magnetized to saturation, e.g., single domains or grains subjected to strong fields which align all domains par-

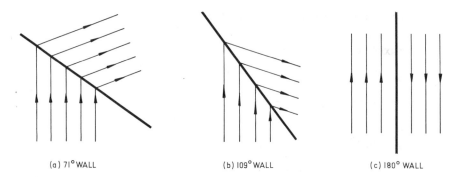

(a) 71° WALL   (b) 109° WALL   (c) 180° WALL

Fig.3.1. Three types of domain wall appearing in magnetite. The angle describing a domain wall is the difference between the directions of magnetization in the domains which it separates, and the wall bisects the included angle, so that there is no net magnetic polarity in its surface.

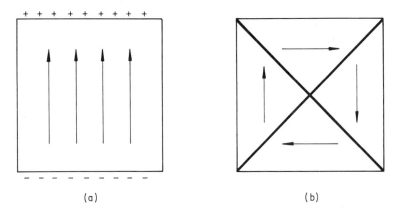

(a)   (b)

Fig.3.2. Possible magnetic structures of a cube, (a) magnetized to saturation, with magnetic polarities on opposite faces and high magnetostatic energy, and (b) divided into four domains by 90° domain walls, leaving no surfaces of magnetic polarity and zero magnetostatic energy, but having domain-wall energy and magnetostrictive strain energy.

allel, producing in effect a large single domain, can be obtained only for ellipsoids of revolution. Thus theories of rock magnetism have generally considered idealized ellipsoidal magnetic grains. Spherical grains, which are the simplest theoretically, suffice for the discussion of several of the most important magnetic properties of all but the smallest grains.

For the general case of a triaxial ellipsoid of volume V, magnetized uniformly with magnetization I per unit volume, magnetostatic energy is:

$$E_m = \frac{1}{2} NI^2 V \tag{3.2}$$

where N is a geometrical parameter which is a function of the dimension ratio of the ellipsoid and the orientation of I with respect to its principle axes, a, b, c. The values of N in the three axial directions are related by the condition:

$$N_a + N_b + N_c = 4\pi \tag{3.3}$$

and for an arbitrary direction of magnetization having direction cosines α_a, α_b, α_c with respect to the axes:

$$N = N_a \alpha_a^2 + N_b \alpha_b^2 + N_c \alpha_c^2 \tag{3.4}$$

The parameter N is termed the demagnetizing or self-demagnetizing factor; it is the coefficient relating the strength of the internal field H_D, arising from the surfaces of magnetic polarity, to the magnetization I. Tensor representation is required for complete generality because H_D is only antiparallel to I if I is parallel to one of the principle axes of the ellipsoid, but a simple way of obtaining a general result is to resolve I along the principle axes to obtain the three components of H_D:

$$\begin{aligned} H_{D_a} &= -N_a I_a \\ H_{D_b} &= -N_b I_b \\ H_{D_c} &= -N_c I_c \end{aligned} \tag{3.5}$$

Then the magnetostatic energy appears as a sum of three integrals:

$$\begin{aligned} \frac{E_m}{V} &= -\int_0^{I_a} H_{D_a} dI_a - \int_0^{I_b} H_{D_b} dI_b - \int_0^{I_c} H_{D_c} dI_c \\ &= \frac{1}{2} N_a I_a^2 + \frac{1}{2} N_b I_b^2 + \frac{1}{2} N_c I_c^2 \end{aligned} \tag{3.6}$$

which is seen to be equivalent to eq.3.2 when the identity 3.4 is noted. The internal field H_D being essentially opposite to the magnetization tends to demagnetize the material. This field is the strongest force causing the sub-division into domains.

A table and graphs of $D = N/4\pi$ for the general case of triaxial ellipsoids are given by Osborn (1945), whose curves give a simple and accurate way of finding the values of D for any dimension ratio. Another tabulation, restricted to ellipsoids of revolution, is by Stoner (1945). A graph of demagnetizing factors vs dimension ratio for ellipsoids of revolution is given as Fig.3.3.

Several special cases can be recognized as applying to particular situations. For example, the case of a sphere is seen by applying the condition of symmetry to eq.3.3:

$$N_{\text{sphere}} = \frac{4}{3}\pi \tag{3.7}$$

Along the axis of a very elongated cylinder $N \to 0$ and across a diameter $N \to 2\pi$. Along the axis of a thin disc $N \to 4\pi$ and along a diameter $N \to 0$. For nearly spherical ellipsoids of principal semi-axes a, b, c, the demagnetizing factors are approximated by:

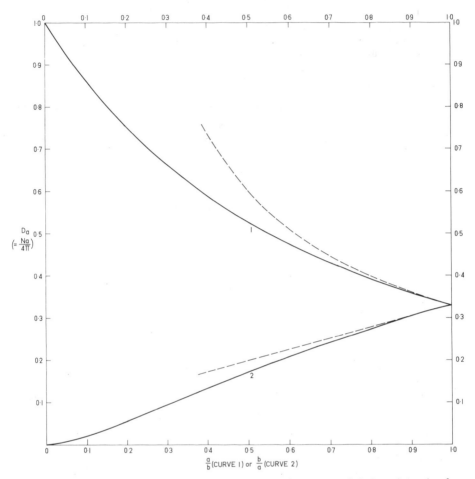

Fig.3.3. Axial demagnetizing factors for ellipsoids of revolution with semi-axes a, b, b. Curve *1* gives data for oblate ellipsoids ($b > a$) and curve *2* for prolate ellipsoids ($a > b$). The dashed lines represent the approximation to D_a by eq.3.8 with $c = b$. Note that by eq.3.3, $D_b = N_b/4\pi = \frac{1}{2}(1 - D_a)$.
(Data for triaxial ellipsoids are given in graphical form by Osborn, 1945.)

$$N_a = \frac{4}{3}\pi\left[1 - \frac{2}{5}\left(2 - \frac{b}{a} - \frac{c}{a}\right)\right] \qquad (3.8)$$

with N_b, N_c given by the same expression with appropriate changes of axes.

The internal field H_I acting on a magnetic material is less than the applied external field H_E by the amount of the self-demagnetizing field. For magnetization along the principal axis of an ellipsoid:

$$H_I = H_E + H_D = H_E - NI \qquad (3.9)$$

Thus a magnetizing field is more effective if applied along a long axis of a body, in which direction the self-demagnetizing factor N is smallest. For this reason an elongated body is more easily magnetized in the direction of its long axis. Not uncommonly highly magnetic basaltic rocks may occur in the form of sheets (as dykes or sills) in which case although the rock is magnetically intrinsically isotropic, its direction of magnetization may be deflected away from the field direction towards the plane of the sheet by the self-demagnetizing field, which operates only normal to the plane of the sheet (section 4.5).

The effect of self-demagnetizing fields in laboratory work is most significant in the measurement of magnetic anisotropy, in which it is important that the observations should give the intrinsic anisotropy of a sample, unaffected by specimen shape. If it were convenient to use spherical specimens there would be no difficulty, but since there is a strong preference for using standard cylindrical cores, the problem is simply one of choosing a length/diameter ratio for the cylinders such that the demagnetizing factor is $\frac{4}{3}\pi$ both along the axis and across a diameter. This problem has no unique solution because even for uniform magnetization the demagnetizing field is not uniform through a cylinder. Then, since susceptibility is field-dependent in a rather complicated way, it follows that the mean effective demagnetizing factor is field-dependent. The only situation for which there is a direct analytical result is for complete magnetic saturation in a very high field; interpolating calculations by Brown (1960), the appropriate length/diameter ratio for this case is 0.91. This agrees reasonably with measurements on nickel cylinders by Porath et al. (1966) who gave ℓ/d ratios of 0.880 to 0.893 for shape isotropy in fields of 10–15 kOe. and 0.902 extrapolated to infinite field. In low fields, H up to 50 Oe, they found:

$$\ell/d = 0.845 + 0.0002 \, H \tag{3.10}$$

These specimens were, of course, much more highly magnetic than any rocks. For work on rocks in high fields it appears that Brown's value ($\ell/d = 0.91$) is the most appropriate. The best low-field ratio is still not finally resolved; measurements on dielectric cylinders by G. J. Tuck appear to be the most relevant. Tuck's limiting value for low susceptibilities is $\ell/d = 0.873$. For most rock magnetic work it is sufficient to cut specimens with $\ell/d = 0.9$.

So far we have considered only bodies which are uniformly magnetized, so that magnetostatic effects can be represented by the single parameter N. The properties of large grains of magnetite and similar strongly magnetic minerals are well accounted for in this way, although the magnetization is not uniform (except in a saturating field) but is merely a net alignment of the domains. Another special case of some interest is the domain structure which terminates at a surface in a regular pattern of alternating polarity, as in Fig.3.4. This situation was analyzed by Kittel (1949), who showed that the magnetostatic energy per unit area of terminating surface for regular arrays of domains of width W was, for lamellar domains (Fig. 3.4, a):

$$E_m = 0.85 \, I_S^2 \, Wf \tag{3.11}$$

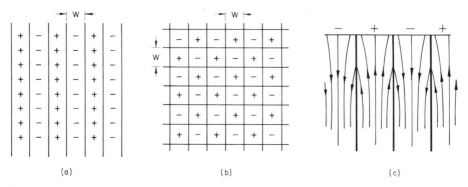

Fig.3.4. Magnetic polarities at a surface terminating (a) lamellar domains and (b) square rod domains magnetized normal to the surface. With the domain width W, as indicated, the surface magnetostatic energies per unit area are given by eq.3.11 to 3.13 derived by Kittel (1949). The factor f in these equations accounts for the deflection of domain magnetization near to the surface, as in (c).

and for square rod domains (Fig.3.4, b):

$$E_m = 0.53\, I_S^2\, Wf \tag{3.12}$$

where f is a factor to allow for diminished surface polarity by deflection of the magnetization within the domains, as in Fig.3.4(c) and:

$$f = \left(1 + \frac{\pi I_S^2}{K_1}\right)^{-1} \tag{3.13}$$

where K_1 is the first magnetocrystalline anisotropy constant, which is considered in sections *1.5* and *3.2*. For magnetite ($I_S = 480$ e.m.u./cm^3, $K_1 = 1.36 \cdot 10^5$ ergs/cm^3), $f = 0.16$. This means that at the surface the domains have a reduced effective intensity $f^{1/2}I_S$ which is $0.39\, I_S$ in magnetite. The domain deflection tendency is thus so strong that the formation of closure domains, as in Fig.3.5, is inevitable. Real domain structures in magnetite

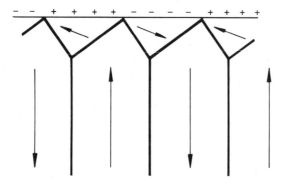

Fig.3.5. Simple closure domain structure at a surface normal to the main domains in a 180° array.

3.2 ANISOTROPY ENERGY

The energy of the spin-orbital coupling considered in section *1.5* imposes an intrinsic anisotropy upon the magnetic alignment of domains. In cubic materials, such as magnetite, the effect at room temperature and above is described by two empirical anisotropy constants, K_1, K_2, so that the energy of domain magnetization in an arbitrary direction with respect to the crystallographic axes can be expressed in terms of the cosines α_1, α_2, α_3 of its direction with respect to the three [100] axes:

$$E_K = K_1(\alpha_1^2\alpha_2^2 + \alpha_2^2\alpha_3^2 + \alpha_3^2\alpha_1^2) + K_2\alpha_1^2\alpha_2^2\alpha_3^2 \tag{3.14}$$

Eq.3.14 satisfies the requirements that it is symmetrical with respect to the three cubic axes and no odd powers of the cosines appear because energies are equal for opposite directions of magnetization. No term of the form $\Sigma\alpha^2$ appears because:

$$\alpha_1^2 + \alpha_2^2 + \alpha_3^2 = 1 \tag{3.15}$$

Terms of the form $\Sigma\alpha^4$, $\Sigma\alpha^6$ and $\Sigma\alpha_p^4\alpha_q^2$ are accounted for by eq.3.14 because, with eq.3.15:

$$\alpha_1^4 + \alpha_2^4 + \alpha_3^4 = 1 - 2(\alpha_1^2\alpha_2^2 + \alpha_2^2\alpha_3^2 + \alpha_3^2\alpha_1^2)$$
$$\alpha_1^6 + \alpha_2^6 + \alpha_3^6 = 1 + 3\alpha_1^2\alpha_2^2\alpha_3^2 - 3(\alpha_1^2\alpha_2^2 + \alpha_2^2\alpha_3^2 + \alpha_3^2\alpha_1^2)$$

and:

$$\alpha_1^4(\alpha_2^2 + \alpha_3^2) + \alpha_2^4(\alpha_1^2 + \alpha_3^2) + \alpha_3^4(\alpha_2^2 + \alpha_1^2)$$
$$= (\alpha_1^4 + \alpha_2^4 + \alpha_3^4) - (\alpha_1^6 + \alpha_2^6 + \alpha_3^6)$$

Higher powers in the α's are not needed (except below room temperature) and even the second term in eq.3.14 can often be omitted without significant error, especially at elevated temperatures.

The energy of magnetization to saturation in any crystallographic direction may be calculated directly from the direction cosines with respect to [100] axes by eq.3.14. The three directions of greatest interest give:

$$E_{100} = 0$$
$$E_{110} = K_1/4 \tag{3.16}$$
$$E_{111} = K_1/3 + K_2/27$$

In magnetite at 290°K, $K_1 = -1.36 \cdot 10^5$ ergs/cm^3 and $K_2 = -0.44 \cdot 10^5$ ergs/cm^3 (see Table 3.I), so that the direction of lowest energy is [111]. The [111] axes are the

TABLE 3.I

Curie points, θ_c, of $(1-x)Fe_3O_4, xFe_2TiO_4$ with values of saturation magnetizations σ_s, magnetocrystalline anisotropy constants K_1, K_2 and magnetostriction constants at 290°K. From data by Akimoto (1962), and Syono (1965)

x	θ_c (°C)	σ_s (e.m.u./g)[1]	K_1 (10^5 erg/cm^3)	K_2 (10^5 erg/cm^3)	λ_{100} (10^{-6})	λ_{111} (10^{-6})
0	575	93	−1.36	−0.44	−20	+78
0.04	550	90	−1.94	−0.18	−6	+87
0.10	520	82	−2.50	+0.48	+4	+96
0.18	470	73	−1.92		+47	+109
0.31	400	59	−1.81		+67	+104
0.56	170	29	−0.70		+170	+92
0.68	80	15	+0.18			

[1] Density $\rho \approx 5.2$ g/cm^3; $I_S = \rho \sigma_s$

crystallographic easy directions in magnetite and [100] axes are the hard directions, that is the domains in magnetite are spontaneously magnetized in (or very near to) [111] directions. The magnetization may be deflected by an applied field; a field of several hundred oersteds must be applied to deflect the magnetization through a [100] direction. As is shown in Chapter 4, the order-of-magnitude of the field required is given by:

$$H_A = \frac{4}{3} \frac{|K_1|}{I_S} \qquad (3.17)$$

which is referred to as the anisotropy field. This gives a rough estimate of the contribution of magnetocrystalline anisotropy to the coercive forces of single domain grains of magnetite. In titanomagnetites at room temperature H_A increases significantly with Ti content (see Table 3.I), being 370 Oe for magnetite but 780 Oe for titanomagnetite with 31% Fe_2TiO_4.

For a simple application of eq.3.14 we may consider the energy of magnetization at an arbitrary angle in a(110) plane, as in Fig.3.6. Putting $\alpha_1 = \cos\phi$, and using (3.15), i.e., $\alpha_2^2 = \alpha_3^2 = \frac{1}{2}(1 - \alpha_1^2)$ for the symmetry of this problem, the anisotropy energy per unit volume is:

$$E_K = \frac{K_1}{4}(4\sin^2\phi - \sin^4\phi) + \frac{K_2}{4}(\sin^2\phi - \sin^6\phi) \qquad (3.18)$$

In magnetite K_1, K_2 are negative and [111] axes are the easy axes so that we are frequently concerned with the deflection of magnetization from a [111] axis by a small angle $\theta = \phi - \cos^{-1}\left(\frac{1}{\sqrt{3}}\right)$. Then to second-order in $\sin\theta$:

$$E_K = \left(\frac{K_1}{3} + \frac{K_2}{27}\right) - \left(\frac{2K_1}{3} + \frac{2K_2}{9}\right)\sin^2\theta \qquad (3.19)$$

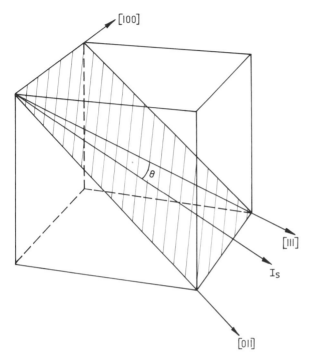

Fig.3.6. Geometry for the calculation of the anisotropy energy of magnetization in an arbitrary direction in a (110) plane (shaded) in a cubic crystal. The energy is given in terms of ϕ, the angle to the nearest [100] axis by eq.3.14, but since the anisotropy energy is negative in magnetite we are more interested in the deflection θ from the nearest easy [111] axis.

The approximation represented by the two terms of (3.19) applies not merely to deflections in a (110) plane but to small deflections in any direction from a [111] axis. This special form of (3.18) is basic to the calculation of the contribution of domain rotation to susceptibility.

3.3 MAGNETOSTRICTIVE STRAIN ENERGY

Eq. 3.14 and the treatment of anisotropy energy up to this point refer to magnetic materials with cubic symmetry, but the cubic symmetry of magnetite is destroyed if an axial stress is applied to it. The magnetic anisotropy is then modified by a superimposed axial term, which is related to the magnetostriction of the material. In fact magnetostriction arises as a strain dependence of the anisotropy and the contribution to magnetic anisotropy energy which arises from the strain is known as magnetostrictive strain energy.

Consider first the simple case of a hypothetical isotropic but magnetostrictive material of positive saturation magnetostriction λ_s, that is its dimension in the direction of saturation magnetization is increased by the fraction $\Delta \ell/\ell = \lambda_s$ relative to the unmagnetized state. In keeping with the assumption that the volume magnetostriction is negligible (a good approximation in most real materials) a lateral contraction $\Delta b/b = -\lambda_s/2$ accompanies the longitudinal extension, in preserving constant total volume during magnetization. (In a material with negative magnetostriction both dimension changes are reversed in sign.) Now suppose that from the same initial unmagnetized state the material is magnetized to saturation with a constant compressive stress σ applied in the direction of the magnetizing field. Then the work done per unit volume of material in producing the magnetostrictive expansion against the stress is $(\lambda_s \sigma)$. This energy must be supplied by the magnetizing field, i.e., there is now an energy of magnetization $(\lambda_s \sigma)$ in the direction of the compression, which has become a hard direction of magnetization. But, in the absence of the applied field the magnetic hardness of the stress axis constrains the spontaneous magnetization to lie in the perpendicular plane, that is, the domain magnetizations are all normal to the stress axis and are not distributed equally in all directions, as was assumed for the initial (ideal) unmagnetized state. Therefore, relative to the assumed (virgin) unmagnetized state there is a magnetostriction $-\lambda_s/2$ accompanying the stress without the application of a field. The magnetostrictive extension resulting from magnetization to saturation with the stress applied is therefore $3\lambda_s/2$, being 1.5 times the extension without stress, and the energy is $\frac{3}{2}\lambda_s \sigma$. If the saturating magnetic field is applied normal to the stress then the magnetostriction in the field direction is $\frac{3}{4}\lambda_s$ but no energy is involved because no change in dimension occurs in the stress direction. We can therefore write the magnetostrictive strain energy due to saturation magnetization at an angle ϕ to the stress in the form of an anisotropy energy:

$$E_\sigma = \frac{3}{2}\lambda_s \sigma \cos^2 \phi \tag{3.20}$$

This is the simplest form of the magnetostrictive strain energy which must be accounted for in domain theory. It applies to any material which has isotropic magnetostriction, such as a rock, in which the crystallographic axes of magnetite grains are randomly oriented. The validity of eq.3.20 for stressed igneous rocks was demonstrated by Stacey (1960a), who showed that by measuring the anisotropy energy of a specimen under controlled stress its magnetostriction is determined directly. This is done by suspending the specimen in a strong field so that it experiences a torque:

$$L = -\frac{dE_\sigma}{d\phi} = 3\lambda_s \sigma \cos\phi \sin\phi = \frac{3}{2}\lambda_s \sigma \sin 2\phi \tag{3.21}$$

A comparison of this equation with some experimental data is shown in Fig.3.7.

In dealing with the domain structures within crystals the assumption of isotropic magnetostriction is no longer adequate. We must calculate the magnetostriction in terms of the orientation of the magnetization with respect to crystallographic axes. To a good

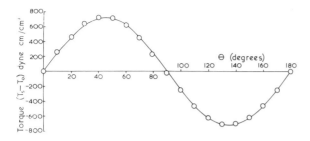

Fig. 3.7. High-field torque curve for a basalt sample subjected to a compressive stress of 500 bar. Data points are differences between values of torque measured with and without stress (a slight intrinsic anisotropy being apparent in the latter case) and the line follows eq.3.21. (After Stacey, 1960a.)

approximation the magnetostriction of a cubic crystal may be represented in terms of an equation analogous to (3.14), in which the two constants λ_{100} and λ_{111} are the magnetostrictions along [100] and [111] axes due to saturation magnetization in those directions. If the direction cosines of the magnetization with respect to [100] axes are $\alpha_1, \alpha_2, \alpha_3$ as before, then the magnetostriction in a direction with cosines $\beta_1, \beta_2, \beta_3$ is:

$$\lambda = -\frac{1}{2}\lambda_{100} + \frac{3}{2}\lambda_{100}(\alpha_1^2\beta_1^2 + \alpha_2^2\beta_2^2 + \alpha_3^2\beta_3^2) + 3\lambda_{111}(\alpha_1\alpha_2\beta_1\beta_2 + \alpha_2\alpha_3\beta_2\beta_3 + \alpha_3\alpha_1\beta_3\beta_1) \quad (3.22)$$

Since our concern is primarily with geometrically simple cases of relative strain between domains magnetized in or close to [111] axes, this equation is frequently simplified. However, the generality is required in all problems concerning magnetostriction within crystals because λ_{100} and λ_{111} have quite different values. Numerical data have been given by Bickford et al. (1955) for magnetite and Syono (1965) for titanomagnetites as functions of temperature. Their room temperature values are reproduced in Table 3.I.

The magnetostriction λ given by (3.22) is relative to the undistorted crystal lattice, so that a compressive stress σ applied in a direction $(\beta_1, \beta_2, \beta_3)$ imposes a magnetostrictive strain energy $(\sigma\lambda)$ on domain magnetization in the direction $(\alpha_1, \alpha_2, \alpha_3)$. No factor 3/2 arises in this case.

Magnetostrictive strain energy (also known as magnetoelastic energy) exercises an influence on domain structure through the mutual strain of adjacent domains. This is an important factor favouring 180° domain walls (Fig.3.1,c) rather than 71° or 109° walls in a domain structure. In particular the four domain structure of Fig. 3.2(b) involves high magnetostrictive strain energy. In magnetite the figure would be a rhomb rather than a cube because the [111] easy directions make 71° and 109° angles with one another, but neglecting this slight error we can readily calculate the strain energy for such a structure. Within the plane of the diagram the magnetostrictive distortion of each domain is opposed by its neighbours but no restraint acts normal to the plane of the figure, in which direction the free transverse magnetostriction, $-\lambda_{111}/2$, acts. This means that all four domains have symmetrical strains $\lambda_{111}/4$ in both directions in the plane of the diagram, relative to the free magnetostrictive strains λ_{111} and $-\lambda_{111}/2$, so that the perpendicular elastic strains are $\pm \frac{3}{4}\lambda_{111}$. Treating the magnetite as elastically isotropic, the corresponding stresses are $\pm \frac{3}{4}\lambda_{111}q/(1+\nu) = \pm(\frac{3}{4}\lambda_{111}\cdot 2\mu)$, where q, ν, μ are Young's modulus, Poisson's ratio, and rigidity. The total strain energy per unit volume is thus:

$$E_\lambda \approx 2 \cdot \tfrac{1}{2}(\tfrac{3}{4}\lambda_{111})^2 \cdot 2\mu = \tfrac{9}{8}\mu\lambda_{111}^2 \qquad (3.23)$$

This energy is reduced by a finer subdivision of the domain structure, as in Fig.3.8, but the increase in domain wall area produces added domain-wall energy, which opposes the change. As in all domain theory calculations the favoured structure can be determined by minimizing the total magnetic energy.

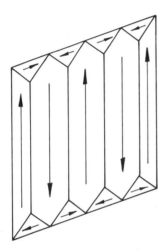

Fig.3.8. Possible domain structure of a rhomb of magnetite in which magnetostrictive strain energy is reduced relative to the structure of Fig.3.2 (b). The main domains are separated by 180° domain walls and strains are restricted to the surfaces with the small closure domains.

Real domain structures are more complicated than the ones considered here and involve flux closure in a three-dimensional domain pattern, but the role of magnetostriction is well illustrated by the foregoing four-domain example. Substitution of numerical values for magnetite ($\mu = 9 \cdot 10^{11}$ dynes/cm^2, $\lambda = 78 \cdot 10^{-6}$) in eq.3.23 gives $E_\lambda = 6 \cdot 10^3$ ergs/cm^3. This is substantially smaller than anisotropy energy, so that magnetostrictive strain is relegated to a secondary role in the determination of domain structure. The most significant effects arise within domain walls.

3.4 DOMAIN WALLS

In eq.1.39 the energy of the exchange interaction between electrons on neighbouring atoms is given as a continuous function of the angle between their spins. If the atoms were remote from all others, then quantization of angular momentum would ensure that the electron spins were either parallel or antiparallel, but in a larger array it is the total angular momentum which is quantized and can only change in units of $\hbar$. Provided the net magnetic moment in any direction of an array of n-coupled spins has one of the permitted values $n\beta, (n-2)\beta, (n-4)\beta, ..., -n\beta$, then the angle between any two may have an arbitrarily small value. For small angles θ, eq.1.39 may be approximated by:

$$E_e = -4s_i s_j A \left(1 - \frac{\theta^2}{2}\right) = E_{e_o} + 2s_i s_j A \theta^2 \tag{3.24}$$

where E_{e_o} is the minimum energy corresponding to parallel spins in the case of positive exchange (A positive). The second term in (3.24) is the excess energy due to spin misalignment. For the purpose of domain theory it is this term which is referred to as the exchange energy.

The important feature of eq.3.24 is the dependence of exchange energy on the square of θ. It means that the total exchange energy of a line of $(n + 1)$ atoms each with electron spins misaligned with its neighbours by an angle π/n, so that the end spins are oppositely directed, as in Fig.3.9, is:

$$E_n = \frac{2\pi^2 s^2 A}{n} \tag{3.25}$$

(a)

(b)

Fig.3.9. (*a*) Progressive rotation of spins through a 180° domain wall. (*b*) An energetically unfavourable alternative spin reversal.

whereas the direct exchange energy between two adjacent atoms with electron spins antiparallel is, by eq.1.34:

$$E_1 = E_e - E_{eo} = 8 s^2 A \qquad (3.26)$$

Thus the exchange energy associated with a domain wall, through which there is a progressive rotation of spins, decreases with the thickness of the wall. However, the total energy of a wall includes also anisotropy energy because the magnetization (spin) vectors within it are deflected out of the [111] easy directions, to which the domains themselves are confined. The wall energy and thickness are calculated by choosing the wall thickness which minimizes the total energy.

We follow here a simplified treatment of this problem by supposing that the variation of spin orientation through a domain wall is in fact linear, as in the approximation represented in Fig.3.9, so that the exchange energy per unit area of the wall is obtained by multiplying eq.3.25 by the number of magnetic atoms per unit area of wall (a^{-2}, where a is the atomic spacing):

$$E_{ew} = \frac{2\pi^2 s^2 A}{na^2} \qquad (3.27)$$

The mean anisotropy energy through a 180°-domain wall is the excess of the average of E_K, as given by eq.3.18, over the easy-direction energy, E_{111}:

$$\bar{E}_K - E_{111} = \frac{1}{\pi} \int_{\phi_0}^{\phi_0 + \pi} \left[\frac{K_1}{4}(4\sin^2\phi - 3\sin^4\phi) + \frac{K_2}{4}(\sin^4\phi - \sin^6\phi) \right] d\phi - \frac{K_1}{3} - \frac{K_2}{27}$$

$$= -0.115 K_1 - 0.021 K_2 \qquad (3.28)$$

and since the volume per unit area of wall is na, the anisotropy energy per unit area of wall is:

$$E_{Kw} = na(-0.115 K_1 - 0.021 K_2) \qquad (3.29)$$

Note that negative K_1, K_2 and [111] easy axes have been assumed, since our concern is primarily with titanomagnetites, and that E_{Kw} is necessarily positive. The total wall energy is thus:

$$E_w = E_{ew} + E_{Kw} \qquad (3.30)$$

and the value of n is determined by the condition that E_w must be a minimum:

$$\frac{dE_w}{dn} = 0 \qquad (3.31)$$

This gives the wall thickness:

$$t = na = \pi s \left(\frac{2A}{a}\right)^{1/2} (-0.115 K_1 - 0.021 K_2)^{-1/2} \approx 10^{-5} \text{ cm} \qquad (3.32)$$

and wall energy:

$$E_w = \pi s \left(\frac{2A}{a}\right)^{1/2} (-0.115\,K_1 - 0.021\,K_2)^{1/2} \approx 1 \text{ erg/cm}^2 \qquad (3.33)$$

The nature of the important approximation made in obtaining this result must be recognized. The anisotropy energy of the wall was assumed to be spread uniformly through it, whereas the balance of exchange and anisotropy energies is maintained at each point in the wall. This means that the angles between adjacent spins are greatest where the anisotropy energy is highest and grade to zero as the spins approach the easy directions of the adjacent domains, as in the curve of Fig.3.10. Since, in titanomagnetites, the [111] (easy)

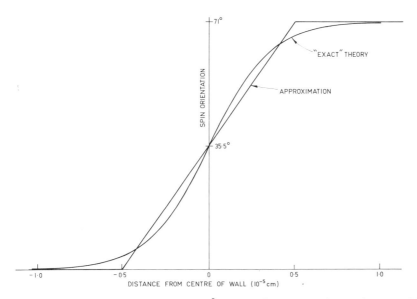

3.10. Variation of spin orientation through a 71° domain wall in magnetite showing the nature of the approximation made in deriving eq.3.32 and 3.33 and the result of a more rigorous calculation, in which the local densities of exchange and anisotropy energies are found to be equal at all points in a wall.

axes make angles of 71° and 109° with each other, it follows that a 180°- domain wall is a combination of 71°- and 109°- walls in which the magnetization turns through an intermediate easy axis. We must therefore enquire what holds a 180°- wall securely as a single domain wall. The simple answer is magnetoelastic energy. If we have 71°- and 109°- walls separated by a wafer of magnetite magnetized almost at right angles to the main domains on either side, then the wafer is subjected to a strong magnetostrictive strain, and bringing the two walls together minimizes the strain energy.

An important feature of the properties of domain walls is that their energies are functions of their positions, by virtue of their interactions with crystal defects. There are two basic kinds of defect which are important in this connection; inclusions and holes interact magnetostatically with domain walls and the stress fields of dislocations interact with the magnetostrictions of walls. Dislocations are extended linear defects responsible for many of the anelastic properties of crystals; there is an extensive literature on dislocations, including monographs by Cottrell (1953) and Friedel (1964).

M. Kersten appears to have been first to note the significance of holes and nonmagnetic inclusions as potential wells for domain walls and suggested that the reduction in wall area, and hence in wall energy, as given by eq.3.33, was responsible for the lower energy of a wall which included a hole. However, Néel (1944) recognized the greater importance of magnetostatic energy in this situation and showed that it is the magnetostatic energy of the hole (or inclusion) which is lowered by the presence of the wall. The shape of the potential well cannot, however, be deduced by comparing the energies of the two situations represented by Fig.3.11 (a and b) because the holes are associated with closure domain structures, as in Fig.3.11,c, causing domain walls to cling to them over a range greater than their own physical dimensions.

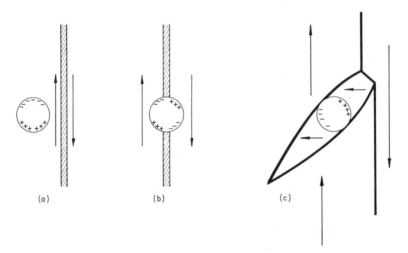

Fig.3.11. Effect of a hole or inclusion on the energy of a domain wall. The magnetostatic energy of the hole in a, separated from a wall, is reduced by a factor rather greater than 2 when the hole is included in the wall. The surfaces of polarity of the hole in situation b closely resemble those of a two-domain grain. When separated from the hole by an external field the wall tends to cling to it by subsidiary closure domains as in c and the energy of this state tends to return the wall to state b. Observations of the effect shown in c confirm the essential validity of Néel's (1944) magnetostatic explanation of the effects of holes and inclusions on coercive force.

The effects of inclusions on the coercive forces of impure magnetites and titanomagnetites are probably very important, particularly so in titanomagnetites which have been partially oxidized, causing exsolution of ilmenite. But in grains of stoichiometric magnetite the greater importance of dislocations has been demonstrated. Parry (1965) measured the coercive forces of dispersed magnetite grains and showed that the grain-size dependence as well as the absolute magnitude of coercivity is affected by annealing. This was shown by Stacey and Wise (1967) to be explicable in terms of the thermal rearrangement of dislocations to a state of partial order (see section 4.1).

A domain wall is normally thin by comparison with the domains which it separates, so that lattice strains in the plane of the wall due to its magnetostriction are prevented by the more massive surrounding material; the magnetostriction appears as a stress pattern in the plane of the wall and a strain normal to it. These interact with the strains and stresses respectively of a dislocation within the plane of the wall. 180°- domain walls in magnetite occur on [110] planes and since the axes of closest packing are [110] axes these must be the orientations of the displacements (Burgers vectors) of the dislocations. A net interaction energy is found to arise only for the case of an edge dislocation with both dislocation axis and Burgers vector in the plane of a wall (Stacey and Wise, 1967), in which case the mutual potential energy of a dislocation and a domain wall are as represented in Fig.3.12. A favourably oriented dislocation is thus seen to act as a potential

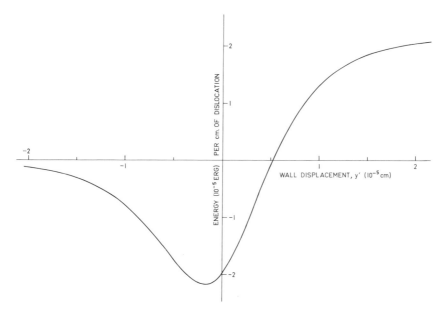

Fig.3.12. Energy of a 180° domain wall per centimetre of an edge dislocation parallel to its plane and distant y' from its centre. The orientation of the magnetization through the wall (at distance y from its centre) is assumed to have the form $\sin\{\theta(y)\} = \tanh\{(y - y')/y_0\}$. Numerical values are appropriate to magnetite.

well for a domain wall (or as a potential barrier for an opposite Burgers vector) but in addition the curve is asymmetrical, i.e., the energy is lower with the wall to one side than the other side of the dislocation. The explanation of coercive force in terms of the superposition of such interactions is considered in section *4.1*.

There are important similarities between domain walls and spin waves, which were mentioned in section *1.2*. Spin waves or magnons are propagating disturbances in the spin structure and are thermally excited. Their mathematical description is very similar to thermal lattice waves or phonons. The important point for the present discussion is that the thermal excitation of the spin structure does not occur as isolated reversals of spins in the parallel, ferromagnetic array, but as progressive spin deflections, which propagate as waves. They are quantized, so that the total magnetic misalignment in a wave is an even multiple of the Bohr magneton. That a spin wave has lower energy and is therefore more easily excited than a single spin reversal is seen from eq.3.25. The longer the wavelength of a spin wave, i.e., the larger the number of nearest neighbour interactions over which the misalignment is spread, the lower the energy, exactly as in the case of a domain wall. This similarity is important in the interaction of spin waves and domain walls. Superposition of a spin wave and a domain wall is equivalent to a transient shift in the wall. Thus spin waves are the mechanism by which domain walls are activated thermally.

3.5 SINGLE DOMAINS AND MULTIDOMAINS

Magnetic properties of mineral grains depend not only upon composition, crystalline perfection, etc., but on the grain size. Although there is no sharp discontinuity in properties at any particular size, the most important distinction is between very small grains, each of which is a single domain and larger grains which are multi-domained. For any material there is a maximum size for spherical single domains, referred to as the

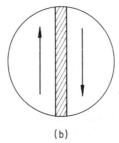

Fig.3.13. Domain structures of a spherical grain of critical size for single-domain structure. (*a*) Single-domain structure. (*b*) Two-domain structure, assuming a domain wall thin compared with grain diameter. This assumption is a poor one in the case of magnetite grains, but nevertheless leads to a critical size in reasonable agreement with observations.

critical size, which can be calculated in several ways with slightly different assumptions but essentially the same result. We follow here the simple approach of Kittel (1949), which considers the two states of a grain of critical size, as represented in Fig.3.13. Smaller grains favour state a and larger grains state b (or even further subdivision), but at the critical size the energies of the two states are equal. In state a the energy is entirely magnetostastic and is given by eq.3.2 with $N = \frac{4}{3}\pi$ and $V = \pi d^3/6$ (for diameter d), so that:

$$E_a = \frac{\pi^2}{9} I_S^2 d^3 = 1.1 I_S^2 d^3 \qquad (3.34)$$

In state b the magnetostatic energy is reduced to $0.46 E_a$, by a calculation due to Néel (1944), but there is an added wall energy ω per unit area over the area $\pi d^2/4$, so that:

$$E_b = 0.5 I_S^2 d^3 + 0.78 d^2 \omega \qquad (3.35)$$

and by equating E_a and E_b we obtain the critical diameter:

$$d_c = 1.3 \frac{\omega}{I_S^2} \qquad (3.36)$$

Substitution of numerical values for magnetite gives $d_c \approx 5 \cdot 10^{-6}$ cm (0.05 μm). Since this is actually thinner than the domain wall thickness by eq.3.32, the calculation is apparently crude. However, a more rigorous calculation, such as applied by Morrish and Yu (1955) to magnetite, in which the whole grain is occupied by a domain wall, essentially recalculates the domain wall energy ω for the special geometry of the grain and arrives at a value not greatly different from that applying to bulk material (~ 1 erg/cm^2). As Amar (1968) pointed out, domain-wall thickness in very fine grains is cramped by a factor of about 2 relative to bulk material.

From measurements of saturation remanence in samples with sub-micron grains of known size distributions, Dunlop (1973a) deduced that the critical size for magnetite grains was between 0.035 μm and 0.05 μm. The most relevant direct observation is by Soffel (1969) who examined domain structures in titanomagnetite grains with a wide range of sizes and constant composition (0.65 Fe$_2$TiO$_4$ · 0.35 Fe$_3$O$_4$) for which the quoted value of I_S was 110 e.m.u. For this material he found the single domain critical size to be about 1 μm. Neglecting the dependence upon composition of ω in eq.3.36, we can see that the critical size for this material is larger than the magnetite value by the ratio of I_S^2, i.e. about 20, which brings the equation to coincidence with Soffel's observation. In fact we expect ω to be smaller than in magnetite because anisotropy energy[1] is smaller (Table 3.I) and allowing for this we arrive at $d_c \approx 0.4$ μm. Since neither the theory nor the observations can give a precise value, the agreement is evidently satisfac-

---

[1] As evidenced by the Curie point, exchange energy is also smaller, but this is not a reliable guide to the relevant exchange energy in a diluted ferrimagnet in which the interactions are antiferromagnetic.

tory. An approximate critical size is in any case sufficient because, as the consideration in the following section shows, there is no sharp demarcation in properties between single domains and small multidomains.

3.6 DOMAIN WALL MOMENTS AND BARKHAUSEN DISCRETENESS

There are two effects which impart permanent magnetic moments to small multi-domain grains, so that they have single-domain as well as multi-domain properties. The term pseudo-single domain grains has been applied to this range. Previous discussions of the problem (Stacey 1961, 1962a, 1963) have been concerned only with the discreteness of domain wall positions (the Barkhausen effect, which is observed in all ferromagnetic materials including magnetite — Domenicali, 1950), but the magnetic moments of domain walls themselves appear at least as important.

The progressive rotation of magnetization through a domain wall produces in a small two-domain grain a situation represented by Fig.3.14. Then assuming that the angle

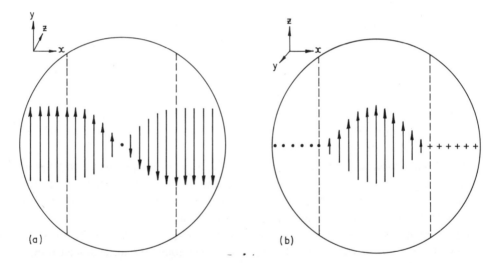

Fig.3.14. Magnetization vectors in a small grain with a 180° wall viewed from two perpendicular directions in the plane of the wall. (a) shows the magnetization components in the directions of the domain magnetizations and (b) normal to the domains. As apparent in (b) there is net magnetic moment perpendicular to the domains.

of the magnetization vector varies linearly through the wall, as in the approximation indicated in Fig.3.10, the vector average moment per unit volume of wall is $2I_S/\pi$ directed normally to the orientations of the domains at the edges of the grain. For a wall of thickness t the moment is $2I_S t/\pi$ per unit area of wall. Thus a small grain, as represented by Fig.3.13, 3.14, has no demagnetized state. When a single 180° domain wall is exactly central in the grain it has a residual (pseudo-single domain) moment normal to the domain magnetizations. In a grain near to the critical size for single domains the wall occupies most of the grain and the pseudo-single domain (p.s.d.) moment may be as much as 50% of the moment of the same grain in the single-domain state. Perhaps it should not be regarded as a two-domain grain, but as a single domain in which the magnetization at the edges has been twisted through 90° or thereabouts. This emphasizes the lack of a sharp discontinuity between single-domain and multidomain properties.

A two-domain grain of diameter d has a p.s.d. moment:

$$m_2 = \frac{2}{\pi} I_S t d^2 \approx 3 \cdot 10^{-3} d^2 \text{ e.m.u.} \qquad (3.37)$$

normal to the domains, numerical values referring to magnetite and d is in centimetres. For higher domain multiplicities the magnetostatic interaction between moments of the several domain walls may tend to align them to produce mutual cancellation, so that the significance of domain wall moments decreases with increasing grain size, quite apart from the $1/d$ dependence of moment per unit volume apparent in eq.3.37. This is consistent with a conclusion by Dunlop (1972) that pseudo-single domain effects are restricted to sub-micron sized magnetite grains, but thermoremanence observations (Chapter 7) indicate that p.s.d. effects are noticeable in grains up to 10 μm diameter.

There is also a p.s.d. moment parallel to one of the domains by virtue of the Barkhausen effect. The crystal defects in a grain and interactions of domain walls with surface irregularities[1] establish minimum energy positions for a domain wall, by the processes considered in section 3.3 and unless one of these minima coincides with the position of the wall which is required to equalize the moments of the opposite domains, the state of zero moment is unstable. This situation is represented by Fig.3.15, in which the domain

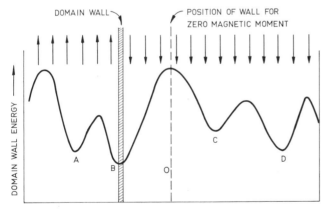

Fig.3.15. Energy of a domain wall as a function of its position in a small grain due to its interaction with crystal defects. The domain wall displacement t' is responsible for the pseudo-single-domain moment of such a grain.

---

[1] The pinning of domain walls by surface defects becomes dominant in small grains.

wall may be found at any of the potential minima A, B, C, D, but not at the zero moment position. The average magnitude of the displacement OB and hence of the p.s.d. moment may be estimated in terms of the widths of the potential wells due to individual crystal defects. In magnetite dislocations appear to be most important and in this case the width of an individual potential well is comparable to the domain wall thickness t (Fig. 3.12). Stacey (1962a) argued that a random superposition of such defects gives an average separation of $t' = \sqrt{3}\,t$ to the potential minima for a domain wall encompassing many defects. As Stacey and Wise (1967) showed the defects cannot be random, but, at least in annealed magnetite, are partially ordered in which case we expect $t < t' < \sqrt{3}\,t$; thus we cannot be seriously in error by assuming:

$$t' = 1.5\,t = 1.5 \cdot 10^{-5} \text{ cm in magnetite} \tag{3.38}$$

The displacement OB may have any value between zero and $t'/2$, so by putting its mean value at $t'/4 = 4 \cdot 10^{-6}$ cm, we can estimate the average p.s.d. moment due to a domain wall of area $A = \pi d^2/4$ in a two-domain grain of diameter d to be:

$$m'_2 \approx 2 I_S A \frac{t'}{4} \approx 0.6 I_S t d^2 \approx 3 \cdot 10^{-3} d^2 \text{ in magnetite} \tag{3.39}$$

The factor 2 in eq.3.39 arises because, by virtue of the displacement by $t'/4$, the magnetization in a volume $(At'/4)$ is *reversed*.

The two contributions to the p.s.d. moment of a two-domain grain, represented by the approximate eq.3.37 and 3.39 are comparable in magnitude but perpendicular to one another and are therefore independent. However, these simple equations apply only to the very small, nearly single-domain grains; for both p.s.d. effects the moments due to several domain walls in a a single grain tend to be mutually cancelling by magnetostatic interaction, so that for larger grains the p.s.d. moments are much smaller than is implied by these equations. The details of the interactions are not adequately understood, but we suppose that the grain surfaces are more or less covered with domain-wall edges which act as semi-independent p.s.d. moments, with dimensions comparable to the thickness of a domain wall (0.1 μm in magnetite: eq.3.32).

We therefore attribute the pseudo-single-domain properties of small multidomains to the termination of domain walls at grain surfaces. Thus we regard every multidomain grain to have, in addition to the multidomain properties arising from its volume, pseudo-single-domain properties arising from its surface and appearing as effective single-domain moments, proportional in number to the surface area, which may be treated as acting independently. The actual number of p.s.d. moments in the surface is some fraction of the possible number, in terms of the available surface area. Since they are associated with domain walls, they must be separated by the small domains (in two directions). Thermoremanence observations, discussed in sections 7.4 and 7.5, indicate that the appropriate fraction is about 1/8, but this depends very closely upon the assumed magnitudes of the moments.

3.7 LAMELLAR INTERGROWTHS

The exsolution of alternate lamellae of magnetite and ilmenite within oxidized grains of titanomagnetite is a common natural occurrence. The lamellar structure is frequently very fine and the extreme dimension ratio of each lamella could allow it to be a single domain if certain conditions are satisfied. Uyeda (1955) examined the problem on the assumption that this is so. It is essential that the plane of the lamella be an easy direction of magnetization; the magnetocrystalline anisotropy may be drastically modified by magnetoelastic energy due to lattice mismatch between the magnetite and ilmenite at the lamellar boundaries (Shive and Butler, 1969), so that this is difficult to determine. Strangway et al. (1968) assumed the magnetic easy directions to remain [111] in magnetite exsolved on (111) planes and were led to an impossible domain structure with internal surfaces of magnetic polarity. However, we suppose the easy direction problem to be overcome and consider the magnetostatic condition for the single-domain structure.

The dimension ratio requirement is complicated by the proximity of adjacent magnetized lamellae, so that a magnetostatic calculation must be applied specifically to an assembly of lamellae and not to each one individually to determine whether they could be single domains. Possible alternative structures are shown in Fig.3.16. The magnetostatic energy in case a has to be compared with the domain wall energy in case b.

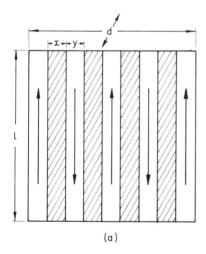
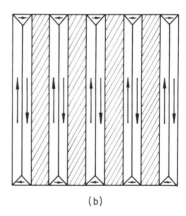

Fig.3.16. Magnetite lamellae alternating with nonmagnetic lamellae; (a) single domains; (b) two-domain lamellae. Which of the states (a) or (b) is favoured depends upon the ratios of the dimensions x, y, l. Dimensions of the grain in the plane perpendicular to l are not significant here, but for convenience in calculations they will both be equated to $d \gg x, y$.

The magnetostatic energy of a grain with the structure of Fig.3.16, (a) may be written in terms of eq.3.11 by subtracting from the energy for domains of width $(y + x)$, the energy which would be due to domains of width x covering a fraction $x/(x+y)$ of the area, which is replaced by nonmagnetic material:

$$E_m = 2\left[0.85\,I_S^2\,(y+x)f - 0.85\,I_S^2\,xf \cdot \frac{x}{y+x}\right]d^2 = 1.7\,I_S^2\,fy \cdot \frac{y+2x}{y+x}d^2 \quad (3.40)$$

The factor 2 takes account of the two faces (top and bottom in the figure) and the saturation magnetization I_S refers to the magnetic component alone. In Fig.3.16 (b) the number of magnetic lamellae is $d/(y+x)$ and each has area (ℓd) so that the domain wall energy is:

$$E_w = \omega\,\ell\,d \cdot \frac{d}{y+x} \quad (3.41)$$

ω being the domain wall energy per unit area. The condition favouring structure a is $E_w > E_m$, i.e.:

$$\ell > 1.7\,\frac{I_S^2\,f}{\omega}\,y\,(y+2x) \quad (3.42)$$

For the case of $x = y$, i.e., equal thicknesses of magnetic and non-magnetic minerals, and substituting numerical values for magnetite as the magnetic mineral, we obtain:

$$\ell > 1.9 \cdot 10^5\,y^2$$

for ℓ, y both in centimetres. This means that for 1 μm grains the lamellae must be thinner than 0.23 μm and for 100 μm grains they must be thinner than 2.3 μm. This condition is by no means always satisfied, but it can be satisfied, so we will now consider briefly the properties of a grain with single-domain lamellae.

The first point to note is that even in 1 μm grains, lamellae of order 0.2 μm thick would have magnetic moments μ such that in all fields H, except carefully annulled laboratory fields, $\mu H/kT \gg 1$, i.e., if the domains acted independently their alignments would not be thermally randomized and T.R.M. in any field would equal the saturation magnetization. However, the magnetostatic energy for parallel magnetizations of all lamellae exceeds the energy for alternating domain polarity by an amount:

$$E_{ms} = \frac{1}{2}\,NI_S^2\,\frac{y^2}{(y+x)^2} \quad (3.43)$$

N being the demagnetizing factor for the whole grain. This energy is equal to the magnetostatic energy of a homogeneous grain with spontaneous magnetization having the average for the heterogenous grain. For magnetite lamellae E_{ms} is so much larger than the other internal energies that the grain behaves essentially as a multidomain. However, there is a difference. Single-domain lamellae cannot exhibit conventional multidomain behaviour because to upset the parallel-antiparallel array of domains (for example in the induction of thermoremanence) a very strong local magnetostatic energy of adjacent parallel domains must be tolerated. If such an effect occurred it would ensure that no thermoremanence could be induced unless a field stronger than some critical value were applied, but no observations of such a phenomenon have been reported. However, except for lamellae very much thinner than those considered above it would be energetically easier to divide one of the lamellae into two domains than to reverse it. Thus there is no way in which the lamellar structure can behave as an array of single domains and for all macroscopic properties we may treat them as multidomains.

Chapter 4

PROPERTIES OF MAGNETITE GRAINS AND OF ROCKS CONTAINING THEM

4.1 COERCIVE FORCE AND GRAIN SIZE

The variation of coercive force (or coercivity – see Fig.1.1), H_c, with grain diameter, d, in granular magnetic materials is quite general and over a wide range of grain sizes is well represented by the relationship:

$$H_c \propto d^{-n} \tag{4.1}$$

Measurements for different materials by several authors have given values of n in the range $0.25 \leqslant n \leqslant 1$. For magnetite grains annealed and dispersed in a much larger volume of non-magnetic material, Parry (1965) obtained $n \approx 0.4$ for d in the range 1.5 μm to almost 80 μm, but the value of n was higher in unannealed grains (see Fig.4.1). That it is

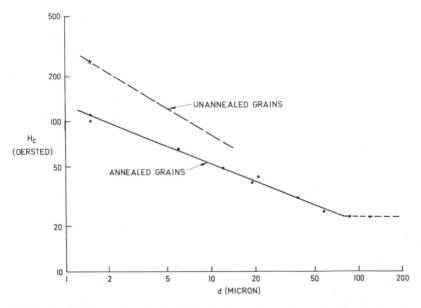

Fig.4.1. Coercive force as a function of grain size for dispersed magnetite powders from the data of Parry (1965). For grains up to about 80 μm in diameter annealed grains follow eq.4.1 with n = 0.4; at larger diameters coercivity becomes less dependent upon grain size because domain structure is not determined primarily by grain size. Parry reported that small unannealed grains gave n = 0.6 with H_c = 250 at d = 1.5 μm. The upper limit for this curve is not clear. In terms of relevance to rock magnetism, these data supersede all measurements on undispersed powders.

not merely the absolute value of coercivity but also its grain size dependence which is affected by annealing is a clue that crystal dislocations are primarily responsible (Stacey and Wise, 1967). Dunlop (1972) has found eq.4.1 to extend into the sub-micron range with $n \approx 0.4$.

Coercivity is a quantitative measure of the resistance of a domain structure to changes. Since we are mainly concerned with multidomain grains, the changes are due to movements of domain walls, which are impeded by crystal imperfections, so that the distribution of the imperfections through a wall is important to the mechanism. We can consider two extreme cases: (*1*) a perfectly ordered array of defects; and (*2*) a randomly distributed array. Then if a wall moves through an ordered array the fluctuations in its energy are directly proportional to the wall area, but if the defects are randomly distributed, so that a wall encompasses an average number n, then the fluctuations in this number are approximately $n^{1/2}$ in magnitude. But since n is proportional to wall area, A, it follows that in this case the fluctuations in wall energy are proportional to $A^{1/2}$. It is the fluctuations which constitute the energy barriers to wall motion so that we may write the average barrier energy as:

$$E_B \propto A^m \tag{4.2}$$

where $m = 1$ for ordered defects, $m = 1/2$ for random defects or some intermediate value for partial order.

Now consider the variation of domain wall area with grain diameter, d. Again we may consider two extremes between which the truth must lie: the domains may extend across the grains as lamellae, in which case $A \propto d^2$, or the domain size may be independent of grain size, i.e., $A \propto d^0$. Thus we may put:

$$A \propto d^l \tag{4.3}$$

with $0 \leqslant l \leqslant 2$.

The energy E_B considered in eq.4.2 is the trough-to-peak energy of the potential barriers impeding a domain wall of area A. We consider a 180° domain wall moving in the x direction under the influence of a field H parallel to the expanding domain; then the energy of the domains due to the field, E_H, varies with x as:

$$\frac{dE_H}{dx} = -2HI_S A \tag{4.4}$$

and the condition that H should be equal to the coercive force, H_c, is that dE_H/dx should be just sufficient to push the domain wall past the steepest point of the potential barrier:

$$-\left(\frac{dE_H}{dx}\right)_{H=H_c} = \left(\frac{dE_B}{dx}\right)_{max} = 2H_c I_S A \tag{4.5}$$

But since the widths of the barriers, as represented by eq.3.38 are independent of their heights:

$$\left(\frac{dE_B}{dx}\right)_{max} \propto E_B \propto A^m \qquad (4.6)$$

and combining eq.4.5, 4.6 and 4.3:

$$H_c \propto \frac{1}{2I_S} A^{m-1} \propto d^{l(m-1)} \qquad (4.7)$$

Thus, Parry's (1965) measurements give:

$$l(1-m) = n = 0.4 \qquad (4.8)$$

for annealed magnetite and a higher value for unannealed grains. We can see immediately that the possibility of an ordered array of crystal defects, which requires $m = 1$, must be discounted because it gives coercivity independent of grain size, regardless of the domain size variation. The other limit, complete disorder ($m = 0.5$), would give $\ell = 0.8$, but since the value of n is larger for unannealed grains, it is necessary to consider higher values of ℓ. The plausible range is indicated in Fig.4.2. If we select a common value of ℓ for both annealed and unannealed grains, then we see that the parameter m, which indicates defect order, increases with annealing quite apart from the general decrease in defect density.

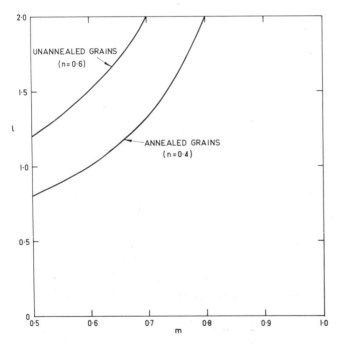

Fig. 4.2. Graphs of domain-size parameter l (eq.4.3) vs defect order parameter m (eq.4.2) for two values of n (eq.4.1), showing the limited ranges of the plausible values of l and m. Low values of m indicate defect disorder and high values of l indicate domain size controlled by grain size.

This was the conclusion of Stacey and Wise (1967); ordering of crystal dislocations is a result of annealing and coercivity is therefore attributed to dislocations, as mentioned in section *3.3*. The preferred value of ℓ is about 1.5, as suggested by the domain theory calculation of Stacey (1959).

Eq.4.1 relates coercive force to grain size for grains in the multidomain range. At small sizes the coercivity must approach the value for single domains and for spherical single domains this is readily calculated from the energy barrier to magnetization rotation given by eq.3.18. The maximum steepnesses of the potential barriers to be crossed between easy directions, with K_1 and K_2 values for magnetite (Table 3.I) are, for opposite directions of rotation:

$$\left| \frac{dE_K}{d\phi} \right|_{max} = \begin{Bmatrix} 1.582 \\ \text{or} \\ 0.654 \end{Bmatrix} \cdot 10^5 \text{ erg/cm}^3 \qquad (4.9)$$

The higher figure represents the barrier for crossing a [100] (hardest) axis and the lower figure a [110] axis. Since rotations of magnetization take the easiest available route the lower figure must be used. To obtain the coercive force, eq.4.9 must be equated to the angular dependence of field energy. If the single domain magnetization I_S makes an angle θ with the field H:

$$\left| \frac{dE_H}{d\theta} \right|_{max} = \left| \frac{d}{d\theta} (H I_S \cos\theta) \right|_{max} = H I_S = 480\, H \text{ erg/cm}^3 \qquad (4.10)$$

Thus a favourably oriented field can just cause magnetization rotation if its strength exceeds $(0.654 \cdot 10^5/480) = 135$ Oe. For an assembly of randomly oriented single domains the corresponding coercivity is about 200 Oe. Higher values are possible with titanomagnetites for which the K_1/I_S ratio is higher, the practical limit being a factor of about 3.

Coercivities exceeding about 200 Oe in magnetite grains require an appeal to shape anisotropy of single-domain grains. For a grain of volume V and principle demagnetizing factors $N_a, N_b, N_b, (N_b > N_a)$, eq.3.2 and 3.4 give the energy of magnetization at an angle θ to the a axis:

$$E_m = \tfrac{1}{2} [N_a + (N_b - N_a) \sin^2\theta] I_S^2 V \qquad (4.11)$$

and the steepness of the energy barrier opposing the reversal of magnetization between easy a axes is:

$$\left| \frac{dE_m}{d\theta} \right|_{max} = \tfrac{1}{2} (N_b - N_a) I_S^2 V \qquad (4.12)$$

Equating this to the field energy dependence (eq.4.10) for volume V gives:

$$H = \tfrac{1}{2} (N_b - N_a) I_S \qquad (4.13)$$

The extreme possible value of $(N_b - N_a)$ is 2π, being appropriate for infinitely long, thin needles, in which case the value is $H = 1500$ Oe. However, this is not observed, because domain rotation in such grains is not coherent. Rather reversed domains are nucleated and the reversals propagate, the barrier energy for this process being much lower (Jacobs and Bean, 1955; Holz, 1970). The maximum observable coercivity remains obscure, but probably does not exceed 400 Oe for magnetite, although perhaps rather more for some titanomagnetites.

Evidence for single domains as the seat of natural remanence in certain rocks is of particular interest because of their very high paleomagnetic stabilities. Evans and McElhinny (1969) investigated very fine magnetite inclusions in the pyroxenes of a South African gabbro, which they concluded were mostly single domains on the basis of the very high alternating fields required to demagnetize them. Since the required demagnetizing field is comparable to the coercivity of remanence and in magnetically stable grains this exceeds the coercive force by a factor greater than 5 (section 4.4), it is misleading to refer to the required demagnetizing field as the coercivity. Nevertheless for magnetite grains demagnetizing fields exceeding 1,000 Oe are evidence of single domains with shape anisotropy. Demagnetizing fields as high as 1,800 Oe, corresponding to coercivities of about 360 Oe, were observed by Evans and McElhinny.

Dunlop (1968) examined the properties of assemblies of single domains in some detail and showed that they were modified by grain interactions. He concluded that independent (non-interacting) single domains occurred rarely, if ever. The interactions compromise the single-domain properties, as is indicated for the case of interacting single domain lamellae in section 3.6. This further emphasizes the lack of a sharp distinction between single domains and multidomains. The problem is considered further in Chapter 7 in connection with thermoremanence.

4.2 SUSCEPTIBILITY OF MAGNETITE GRAINS

When a grain is magnetized with intensity I by an externally applied field H_e, the effective internal field H_i acting on the material of the grain is less than H_e by an amount $(-NI)$, the self-demagnetizing field arising from the surfaces of magnetic polarity, as discussed in section 3.1:

$$H_i = H_e - NI \tag{4.14}$$

But the intrinsic susceptibility of the material of the grain χ_i relates I to H_i, for small fields:

$$I = \chi_i H_i = \chi_i (H_e - NI) \tag{4.15}$$

so that the observed volume susceptibility of the grain, which relates I to H_e is:

$$\chi = \frac{I}{H_e} = \frac{\chi_i}{1 + N\chi_i} \tag{4.16}$$

SUSCEPTIBILITY OF MAGNETITE GRAINS

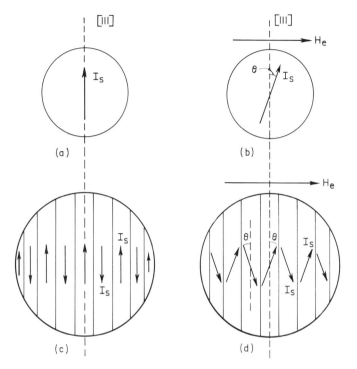

Fig.4.3. Deflection of magnetization I_S in a single domain, (a) and (b), and a simple multidomain grain, (c) and (d), by a perpendicular field. In both cases the intrinsic susceptibility is due to the component of magnetization in the direction of the field, $I_S \sin \theta$. Note that in the multidomain case no surfaces of magnetic polarity appear at the domain walls because θ is the same for domains of both polarities.

The intrinsic susceptibility χ_i is the susceptibility which would in principle be observed if it were possible to examine the material of the grain in bulk in a permeameter which avoided self-demagnetizing effects. The susceptibility χ, which is always observed in rock magnetism, turns out to provide rather a poor measure of χ_i for strongly magnetic minerals such as magnetite, because $\chi_i > 1/N$, so that $\chi \approx 1/N$. Eq.4.16 refers to the susceptibility per unit volume of magnetic material. Applied to a rock with a volume fraction $f \ll 1$ of magnetic mineral, this gives:

$$\chi = \frac{f\chi_i}{1 + N\chi_i} \qquad (4.17)$$

if N is the same for all magnetic grains. Stacey (1963) suggested that for an assembly of randomly oriented ellipsoidal grains of dimension ratio 1.5 : 1, which is typical of the larger titanomagnetite grains found in rocks, the appropriate average value of N is 3.9.

The intrinsic susceptibility is a structure-sensitive parameter which shows a general inverse correlation with coercivity over a very wide range of synthetic ferromagnetic materials (Kittel, 1949). The data plotted by Kittel suggest a general correlation of the form:

$$\chi_i H_c \approx \text{constant} \tag{4.18}$$

and we can expect a similar relationship to apply to the ferrimagnetic minerals.

There are two distinguishable processes contributing to χ_i, domain rotation and domain-wall translation, although normally the cannot be observed separately. Domain rotation occurs when a field is applied normal to a domain causing the magnetization to turn against the force of magnetocrystalline anisotropy. The low field susceptibility resulting from this process, $\chi_\perp$, may be regarded as a constant of the material and applies equally to single domains and multidomains, as indicated in Fig. 4.3. The angle of deflection θ is calculated by minimizing with respect to θ the sum of two energies, the energy E_H of the induced moment $I_S \sin \theta$ in the field and the anisotropy energy E_K due to deflection of the magnetization away from the easy [111] axis:

$$E_H = -H_i I_S \sin \theta \tag{4.19}$$

$$E_K = -\tfrac{2}{3}(K_1 + \tfrac{1}{3} K_2) \sin^2 \theta \tag{4.20}$$

Eq. 4.20 is the second (θ dependent) term of eq. 3.19. The condition $(d/d\theta)(E_H + E_K) = 0$ gives:

$$\chi_\perp = \frac{I_S \sin \theta}{H_i} = \frac{I_S^2}{-\tfrac{4}{3}(K_1 + \tfrac{1}{3} K_2)} \tag{4.21}$$

$$= 1 \cdot 1_6 \text{ e.m.u. for magnetite.}$$

This is a characteristic constant for the material. It enters susceptibility calculations, but is not directly observable in either multidomains or single domains because eq. 4.16 applies directly to the multidomain situation with $\chi_i \neq \chi_\perp$ and single domains are normally elongated, so that shape anisotropy exercises a stronger control on susceptibility than the crystalline anisotropy.

For non-spherical single domains of dimension ratio $a : b : b\ (a > b)$, the easy axis is the a axis and the shape anisotropy energy arising magnetostatically for magnetization at an angle θ to the a axis is:

$$E_m = \tfrac{1}{2}(N_b - N_a) I_S^2 \sin^2 \theta \tag{4.22}$$

If the deflection is due to a field H_e applied normal to the a axis, then the magnetic field energy is:

$$E_H = -H_e I_S \sin \theta \tag{4.23}$$

and θ is determined by minimizing $(E_m + E_H)$ with respect to θ, giving the susceptibility normal to the easy direction for single domains with shape anisotropy:

$$\chi'_\perp = \frac{I_S \sin\theta}{H} = \frac{1}{N_b - N_a} \tag{4.24}$$

This applies, rather than eq.4.21, if E_m by eq.4.22 exceeds E_k by eq.4.20. This condition is satisfied if $(D_b - D_a) = (N_b - N_a)/4\pi > 0.069$ for magnetite and from the data of Fig. 3.3 this requires a dimension ratio $a/b > 1.2$, a condition which is normally satisfied. For an assembly of prolate single domains with randomly oriented axes:

$$\chi = \frac{2}{3} \chi'_\perp \tag{4.25}$$

The lower limit of $\chi_\perp$ is given by $(N_b \to 2\pi, N_a \to 0)$ for very elongated grains, in which case $\chi \approx 0.1$ e.m.u.

In bulk ferromagnetics, and particularly in magnetically soft materials, low field susceptibility is dominated by domain-wall movements. The simplest case is the application of a field parallel to one domain which thus expands relative to its oppositely magnetized neighbour by motion of the 180° wall separating them. For this reason the resulting susceptibility is represented by $\chi_{//}$, indicating intrinsic susceptibility parallel to the domains. It is this particular susceptibility which is most obviously related to coercivity H_c, since $\chi_{//}$ measures the ease and H_c the difficulty of domain-wall movement.

Following Stacey (1963), we consider a simple model of the sequence of energy minima and maxima of a wall of area A, by representing the energy as a sinusoidal function of its position x (normal to A), with the addition of field energy due to a field (internal) H applied parallel to the domain at low x:

$$E = \frac{E_0}{2}\left[1 - \cos\left(\frac{2\pi x}{t'}\right)\right] - 2AI_S Hx \tag{4.26}$$

where E_0 represents the potential barrier height and t' is the separation of the minima, as in section 3.3. The coercive force is obtained by putting $(dE/dx)_{max} = 0$, because the field is then just strong enough to impel the wall past the barriers (gradient of field energy equal and opposite to the maximum gradient of barrier energy). This gives:

$$H_c = \frac{\pi}{2} \frac{E_0}{At' I_S} \tag{4.27}$$

For small fields the displacement of the wall x from a potential minimum (at $x = 0$) is small enough to simplify (4.26):

$$E = \pi^2 E_0 \frac{x^2}{t'^2} - 2AI_S Hx \tag{4.28}$$

and the equilibrium condition $(dE/dx = 0)$ gives:

$$x = \frac{At' I_S H}{\pi^2 E_0} \tag{4.29}$$

so that the induced moment is:

$$M = 2AI_S x = \frac{2A^2 t'^2 I_S^2 H}{\pi^2 E_0} \tag{4.30}$$

If there are n/v favourably oriented domain walls per unit volume of magnetic material, then:

$$\chi_{\parallel} = \frac{n}{v}\frac{M}{H} = \frac{2nA^2 t'^2 I_S^2}{v\pi^2 E_0} \tag{4.31}$$

and, combining (4.27) and (4.31):

$$\chi_{\parallel} H_c = \frac{n}{v} At' I_S/\pi \tag{4.32}$$

For a rough numerical substitution we may consider a grain of diameter 0.5 μm which is the mean size appropriate to the two domain structures in magnetite (by the condition that the surface area should be about 8 times the cross-section of one pseudo-single domain volume, as discussed in section *3.5*). Then $n/v = 1/d^3$, $A = d^2$, $t = 0.15$ μm, giving a numerical value for the constant in eq. 4.18:

$$\chi_{\parallel} H_c \approx 45 \text{ e.m.u.} \tag{4.33}$$

The approximate nature of this result should be emphasized, because in a soft but polycrystalline magnetic material, in which the domains may be regarded as randomly oriented, we can consider χ_i to be dominated by $\chi_{\parallel}$ which is much greater than $\chi_{\perp}$, so that $\chi_i \approx \chi_{\parallel}/3$ and $\chi_i H_c \approx 15$ e.m.u. However in rocks $\chi_{\parallel}$ does not necessarily exceed $\chi_{\perp}$ as may be seen by substituting numerical values of H_c in (4.33) from the data of Fig. 4.1. At $d = 1$ μm we obtain $\chi_{\parallel} = 0.35$ e.m.u. and at $d = 50$ μm, $\chi_{\parallel} = 1.6$ e.m.u. The break-even point, $\chi_{\parallel} = \chi_{\perp} = 1.16$ e.m.u. corresponds to a coercivity of 39 Oe and a grain size of 20 μm. This is interesting because it has for some time been realized that there is a discontinuity in magnetic properties at about this grain size although not attributed to this effect (Parry, 1965; Dickson et al., 1966).

The possibility of observing $\chi_{\parallel}$ and $\chi_{\perp}$ semi-independently by domain alignment experiments is considered in section *4.3*. The general conclusion is that such experiments have not given the expected results, that the domain structures of magnetite grains are more complicated than had been envisaged and in particular that 71° domain wall movements account for most of the apparent domain rotations which are observed. Thus, in practice $\chi_{\parallel}$ and $\chi_{\perp}$ are not clearly distinguishable and all that we observe is an appropriate average. The obvious average to take for an isotropic rock is:

$$\chi_i = \frac{1}{3}\chi_{\parallel} + \frac{2}{3}\chi_{\perp} \tag{4.34}$$

so that the range of χ_i is quite limited over the range of grain sizes which are of greatest interest. For 1 μm grains we obtain $\chi_i = 0.89$ e.m.u. and for 50 μm grains $\chi_i = 1.31$ e.m.u., corresponding to $\chi = 0.199 f$ e.m.u. and $0.214 f$ e.m.u., respectively, by eq. 4.17. These values are in accord with data on dispersed magnetites and titanomagnetites by several authors, including Parry (1965) and Bhathal and Stacey (1969).

It is evident that if susceptibilities are observed to depart significantly from the range $0.20 f$ to $0.215 f$ the explanation is to be sought in a more extreme dimension ratio for the grains, causing a change in $\bar{N}$ from the value of 3.9 assumed, or else, of course an error in the determination of f. Conversely, if there is no reason to doubt the normality of magnetic grain shapes and assuming that the only magnetic mineral is magnetite, its volume fraction ($f \ll 1$) can be estimated directly from susceptibility.

4.3 ANISOTROPY IN MAGNETITE-BEARING ROCKS

That magnetic anisotropy is associated with the layered fabric of sedimentary rocks has been known for many years but it was not until the development of interest in the fabrics of less obviously anisotropic rocks that serious effort was devoted to understanding magnetic anisotropy of geological materials. An early abstract by Graham (1954) drew attention to the possible usefulness of magnetic anisotropy in studying rock fabrics, but this was not pursued until a new stimulus to the subject was provided by paleomagnetism. In this case there was a need to know whether anisotropy affected the directions of natural magnetizations of rocks, so that measurements were made on rocks previously regarded as completely isotropic (e.g., Stacey, 1960b). They confirmed Graham's conclusion that magnetic anisotropy could be observed in almost any rock and that it was by far the most sensitive petrofabric indicator then discovered.

Graham (1954) observed anisotropies by measuring susceptibilities in various directions in rock samples. This is a direct and convincing method which has been used by a number of authors, but lacks the sensitivity of the torque-meter methods, in which specimens are suspended in magnetic fields so that they tend to turn to align their axes of easiest magnetization, or highest susceptibility, with the fields. There are two types of torque-meter in use; they are referred to as low field and high field instruments. The first low field torquemeter appears to have been due to G. Ising, whose basic idea was developed and refined by King and Rees (1962). It consists essentially of a specimen holder supported by a fine suspension between a pair of Helmholtz coils. The coils are mounted on a turntable and scale with their axis horizontal, so that an alternating field can be applied to specimens in a series of known horizontal directions, in each of which the resulting torque is measured in terms of the twist of the suspension needed to restore the specimen to its zero field orientation. Alternating fields (50 or 60 Hz) are used in low field torque meters so that no torque results from magnetic remanence in specimens.

The high field torque-meter method, well known in the measurement of magnetocrystalline anisotropy in solid state physics (Bozorth, 1951), was first applied to rocks by Stacey (1960b). Its advantage is that with high fields it is possible to observe crystallographic alignment of cubic minerals, as well as the grain shape and uniaxial crystal alignments which are observed by low-field methods. For absolute measurements steady fields exceeding 10^4 Oe must be used, as the measured torque arises from the angular dependence of the energy of saturation magnetization.

Anisotropy of magnetite-bearing rocks is due almost entirely to alignment of the elongations of the magnetite grains. Consider the simple case of a rock with a volume fraction f of identical ellipsoidal grains of semi-axes a, b, b and principal demagnetizing factors N_a, N_b, N_b, all aligned the same way and having an isotropic intrinsic susceptibility χ_i. Its susceptibilities in the a and b directions are:

$$\chi_a = \frac{f\chi_i}{1 + N_a \chi_i} \tag{4.35}$$

$$\chi_b = \frac{f\chi_i}{1 + N_b \chi_i} \tag{4.36}$$

When a small field H is applied at an angle θ to the a axis, the a and b components of induced magnetization are $(\chi_a H \cos\theta)$ and $(\chi_b H \sin\theta)$ and these can be resolved parallel and perpendicular to H:

$$I_{\parallel} = (\chi_a \cos^2\theta + \chi_b \sin^2\theta) H = [\tfrac{1}{2}(\chi_a + \chi_b) + \tfrac{1}{2}(\chi_a - \chi_b)\cos 2\theta] H \tag{4.37}$$

$$I_{\perp} = (\chi_a - \chi_b) H \sin\theta \cos\theta = \tfrac{1}{2}(\chi_a - \chi_b) H \sin 2\theta \tag{4.38}$$

Thus by virtue of the shape anisotropy of the grains the magnetization is not parallel to the field; a perpendicular component $I_\perp$ appears. Measurements of susceptibility anisotropy by the original Graham method determine the angular dependence of susceptibility, $\chi_{\parallel} = I_{\parallel}/H$, the limitation being that if the anisotropy $(\chi_a - \chi_b)/(\chi_a + \chi_b)$ is small, very precise observations are required. On the other hand the low field torque-meter measures the torque L (per unit volume of the specimen):

$$L = I_\perp H = \tfrac{1}{2}(\chi_a - \chi_b) H^2 \sin 2\theta \tag{4.39}$$

and this is directly proportional to the susceptibility difference. It is also important that since torque is proportional to H^2 the polarity of H is not significant and an alternating field of frequency high compared with the period of the suspension system gives a torque proportional to $\overline{H^2}$. The dependence upon H^2 has another advantage — it gives a very wide range of instrumental sensitivity since a 100 : 1 range in Helmholtz coil current gives a 10^4 : 1 sensitivity range. The simplicity and versatility of the low field torque method have made it the preferred one for most anisotropy measurements.

Interpretation of susceptibility anisotropy in terms of the grain ellipticities requires an estimate of $(N_b - N_a)$ and hence of N_b, N_a from $(\chi_a - \chi_b)$ and necessitates an absolute susceptibility measurement as well as the difference measurement:

$$\chi_a - \chi_b = \frac{f\chi_i}{1 + N_a \chi_i} - \frac{f\chi_i}{1 + N_b \chi_i} = \frac{f\chi_i^2 (N_b - N_a)}{(1 + N_a \chi_i)(1 + N_b \chi_i)} \tag{4.40}$$

so that:

$$(N_b - N_a) = \frac{f(\chi_a - \chi_b)}{\chi_a \chi_b} \approx 5 \frac{\Delta\chi}{\chi} \qquad (4.41)$$

for $\Delta\chi/\chi \ll 1$, since $\chi \simeq 0.2 f$ by the argument of section *3.2*.

Now consider the same specimen subjected to a steady field H, high enough to produce saturation magnetization in the field direction independently of its orientation with respect to the field. The sum of its magnetostatic and field energies per unit volume is:

$$E = f(\tfrac{1}{2} N I_S^2 - H I_S) \qquad (4.42)$$

where $N = (N_a \cos^2\theta + N_b \sin^2\theta)$, as shown in section *3.1*. The torque (per unit volume) which it experiences, is, therefore:

$$L = -\frac{dE}{d\theta} = -\frac{f}{2} I_S^2 \frac{dN}{d\theta} = -\frac{f}{2} I_S^2 (N_b - N_a) \sin 2\theta \qquad (4.43)$$

To determine $(N_b - N_a)$ both f and I_S, the spontaneous magnetization of the mineral, must be known. For a simple titanomagnetite I_S may be estimated from the Curie point (and hence composition) of the mineral and (fI_S) is the saturation magnetization of the whole sample.

So far it has been assumed that all of the mineral grains are similar, oriented ellipsoids. The real situation in a rock is that the mineral grains are of different shapes, sizes and orientations, but there is a preferred orientation of long axes. The grains may then be treated in terms of the equivalent ellipsoid, that is the ellipsoid which would have the same anisotropy. The irregularities of grain shape normally average out resulting in a torque curve of $\sin 2\theta$ form as in the above equations. Generally the magnetic grains in igneous rock have axial ratios of about 1.5 : 1 but the preferred alignment is slight, so that the dimension ratio, $a/b : 1$, of the equivalent ellipsoid is much closer to unity. We may then refer to a fractional alignment, $(a/b - 1)/0.5$, which is zero for a perfectly isotropic rock and unity for a rock with complete grain alignment.

Crystalline alignment of cubic minerals in a specimen is apparent in a high field measurement as a $\sin 4\theta$ term in the torque curve. This is a consequence of the variation of the magnetocrystalline anisotropy energy within any plane in a single crystal, as in eq.3.18. Such $\sin 4\theta$ terms cannot in principle be observed in low field measurements because the 4θ terms depend upon non-linear magnetization, i.e., an approach to saturation. Stacey (1960b) sought evidence of crystalline alignment in a series of fourteen specimens cut from a single specimen of Tasmanian dolerite, but found only randomly oriented $\sin 4\theta$ terms in the torque curves, of magnitudes consistent with the statistical alignment of about $n^{1/2}$ grains in each n-grained sample. However, the $\sin 2\theta$ terms in the torque curves had a constant orientation with a standard deviation less than $8°$ which indicated a small (4%) fractional alignment of the grain elongations. This result indicated that the alignment of grain elongations was not accompanied by an alignment of crystal axes of the cubic magnetic minerals and therefore that the crystal axes show no preference for alignment with the grain axes. Thus, in measuring the grain alignment we are entitled to assume, as in eq.4.30 to 4.43, that the magnetic material is intrinsically isotropic.

Since the statistical "noise" level of anisotropy measurements is represented by the alignment of $n^{1/2}$ grains in an n-grained sample, i.e., a fractional alignment of $n^{-1/2}$, it follows that the corresponding "noise" level for the dimension ratio of the equivalent ellipsoid is $a/b = (1 + 0.6\ n^{-1/2})$. This effect can be noticeable at 1% fractional alignment. With coarse grained rocks which are only weakly anisotropic it is necessary to use specimens as large as possible to increase n. Lower limits to measurable anisotropy in specimens of volume 6 cm³ for different grain sizes and susceptibilities are shown in Fig.4.4.

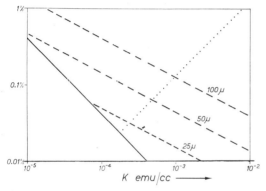

Fig.4.4. Lower limits to measurable anisotropy in specimens of volume 6 cm³ set by instrumental sensitivity (full line) and statistical anisotropy for magnetite grains of the indicated sizes (broken lines). The dotted line indicates the level below which specimen shape becomes significant.
(Figure reproduced by permission, from King, 1966.)

Since there are two basic mechanisms, domain rotation and domain wall motion, which are responsible for the contributions $\chi_\perp$ and $\chi_{//}$ to intrinsic susceptibility it follows that an anisotropy of susceptibility must result from a non-random distribution of domain orientations in a rock. Stacey (1961, 1963) suggested that this effect could be observed as an induced anisotropy of susceptibility after subjecting a specimen to a strong alternating field, an increase in susceptibility being expected in the direction in which the field was applied and a decrease in the transverse directions. Bhathal and Stacey (1969) verified the prediction qualitatively but found the effect to be more complicated than had been anticipated. They carried out a series of experiments with specimens having magnetite grains with aligned axes and found that for the larger grain sizes the induced anisotropy was much greater when the field had been applied along the intrinsic easy axis than when the field had been applied perpendicular to the easy axis. In some cases the difference was greater than the factor 2 permitted by the hypothesis that each grain had a domain structure such as Fig.3.8 which was simply reoriented to the nearest [111] crystal axis. A further important feature of the observations was that the induced anisotropy appeared in fields much smaller than the coercive force and saturated in fields as low as 50 Oe which have hardly any influence on stable thermoremanence. The essential conclusion is that except for very small multidomains the domain structure is more complicated than the simple model of Fig.3.8. and that domain rotation effects proceed largely by 71° domain wall movements. This invalidates the supposition that individual grains have domain anisotropy of the type implied by Fig.3.8 and ensures that for most purposes the intrinsic susceptibilities of magnetite grains should be regarded as isotropic. This is important to the theory of multidomain thermoremanence (Bhathal et al., 1969 – see Chapter 7).

Graham's (1954) original idea to use magnetic anisotropy as a petrofabric tool in structural geology has made some progress but is still far from wide adoption. The correlation of magnetic anisotropy with established fabric is well documented by numerous authors (e.g., Balsley and Buddington, 1960; Stacey et al., 1960); there are still rather few reports of its use to determine fabric where this was not otherwise apparent (Brown et al., 1964; King, 1966).

4.4 SATURATION REMANENCE AND COERCIVITY OF REMANENCE

Fig.4.5 is a plot of the intrinsic hysteresis loop of a strongly magnetic material, that is a plot of magnetization I against the internal field acting on the material. When the material is in the form of a grain of demagnetizing factor N then the internal field H_i is related to the externally applied field H_e by eq.3.9, $H_i = (H_e - NI)$. Since it is the variation of I with H_e which is observed we wish to derive the observed $I - H_e$ loop from the intrinsic $I - H_i$ loop or vice versa. A simple geometrical way of doing this, due to Néel (1955), is to draw on the intrinsic plot a new set of inclined axes for I with a gradient of $-1/N$, such as the broken lines in Fig.4.5, and then the extrinsic loop can be read from intercepts with the inclined axes.

The simplest illustration of this process is to consider the material to follow the path AB from saturation at A to removal of the external field H_e at B, as indicated by the fact that the axis of gradient $-1/N$ through the origin (O) intercepts the loop at B. The external field having been removed, the internal field is simply $-NI_{RS}$ where I_{RS} is the saturation remanent magnetization remaining at B. But the maximum reverse internal field which the material can withstand is very nearly equal to its coercive force, H_c, which is the field at point C on the loop, at which the magnetization is reduced to zero. Except for very fine grains the gradient (dI/dH) of the section of loop BCD is very large in a material such as magnetite, the internal field being almost the same at all three points. Thus:

$$-H_c \approx -NI_{RS} \quad ; \quad I_{RS} = H_c/N \qquad (4.44)$$

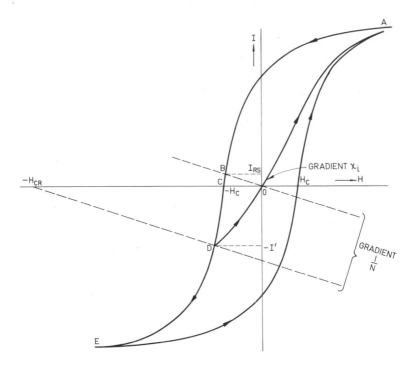

Fig.4.5. Hysteresis loop for a magnetic grain of demagnetizing factor N, showing how saturation remanence, I_{RS}, coercivity, H_c, and coercivity of remanence, H_{cR}, are related. χ_i is the intrinsic susceptibility.

as was shown by Néel (1955). Thus saturation remanence is proportional to coercive force, the constant of proportionality being very close to $1/N$, and for magnetite grains in igneous rocks $\bar{N} = 3.9$. In column 3 of Table 4.I values of I_{RS} calculated from Parry's (1965) H_c data are listed for comparison with the observed values of I_{RS} in column 4. Although the correspondence between theoretical and observed values is good, the theoretical values are consistently low by a factor averaging 0.88, whereas by virtue of the neglect of the slope of the line BC the theoretical values should be slightly too high. The reason is simply that measurements of coercive force of assemblies of grains yield values biassed towards the magnetically softer grains, whose reversed magnetization cancels the positive magnetization of the more coercive grains. In fact we suggest that values of coercivity deduced from saturation remanence would be more representative of assemblies of grains than are direct measurements from a hysteresis loop.

Continuing the demagnetization loop in Fig.4.5 to the point C by applying a reversed field we have reduced the magnetization I to zero. The internal and external fields are thus equal; the externally applied field at $I = 0$ is equal to the intrinsic coercive force H_c of the grain (subject of course to the averaging problem of the previous paragraph).

SATURATION REMANENCE AND COERCIVITY OF REMANENCE 81

TABLE 4.1

Numerical relationships between magnetic properties. Comparison of theory with data by Parry (1965)

1 Grain diameter, d^* (μm)	2 Coercivity, H_c^* (Oe)	3 $H_c/N = I_{RS}$ (e.m.u.) ($N = 3.9$)	4 Saturation remanence I_{RS}^* (e.m.u.)	5 Coercivity of remanence H_{CR}^* (Oe)	6 $(H_{CR}-H_c)/3.9$ $= \chi_i H_c$ (e.m.u.)	7 $(H_{CR}/H_c - 1)/3.9$ $= \chi_i$ (e.m.u.)	8 $(1-H_c/H_{CR})/3.9$ $= \chi$ (e.m.u.)	9 Susceptibility, χ^* (e.m.u.)
120	23	5.9	7.3	170	38	1.64	0.22_2	0.24
88	23	5.9	7.7	180	40	1.75	0.22_4	0.21
58	25	6.4	7.6	190	42	1.69	0.22_3	0.22
37	31	7.9	9.6	200	43	1.40	0.21_7	0.22
21	43	11.0	9.0	230	46	1.06	0.19_9	0.21
19	39	10.0	12.0	220	46	1.19	0.21_1	0.19
12	49	12.6	14.0	230	46	0.95	0.20_2	—
6	66	16.9	21.0	230	42	0.64	0.18_3	0.19
1.5	100	25.6	30.3	320	45	0.45	0.14_0	0.16
1.5	110	28.2	30.0	320	43	0.39	0.13_4	0.19

* Asterisked parameters represent Parry's measurements on magnetite grains; the others are calculated in the manner indicated. Values of remanence and susceptibility are normalized to refer to pure (undiluted) magnetic mineral.

Coercivity of remanence is the strength of the reversed field which must be applied to bring the material to point D on the loop, from which it would return by path DO to the demagnetized state if the field were removed. Let the reversed magnetization at D be $-I$, and the internal field $-H_{iD}$ then the externally applied reversed field at D is $-H_{cR}$, the negative of the coercivity of remanence, and:

$$-H_{iD} = -H_{cR} - N(-I') \tag{4.45}$$

But to a good approximation $H_{iD} = H_c$ and since the gradient of the line DO is χ_i, the intrinsic susceptibility at low fields (i.e., through the origin), we also have $I' = \chi_i H_c$, so that:

$$H_{cR} = H_c (1 + N\chi_i) \tag{4.46}$$

This leads immediately to a check on eq.4.33 because we can rewrite eq.4.46:

$$\chi_i H_c = (H_{cR} - H_c)/N \tag{4.47}$$

which is estimated in section *4.2* to have an approximately constant value of 45 e.m.u. Column 6 of Table 4.I lists the values of $\chi_i H_c$ determined from Parry's data on H_c and H_{cR}, which are in satisfactory accord with the theoretical value.

There are two more relationships of interest which follow directly from the above equations and eq.4.16:

$$\chi_i = \left(\frac{H_{cR}}{H_c} - 1\right)/N \tag{4.48}$$

and:

$$\chi = \frac{\chi_i}{1 + N\chi_i} = \left(1 - \frac{H_c}{H_{cR}}\right)/N \tag{4.49}$$

Values of intrinsic susceptibility calculated from (4.48), listed in column 7 of Table 4.I, cover the range expected from the arguments in section *4.2*, and calculated values of extrinsic susceptibility by eq.4.49 (column 8) compare well with Parry's measurements (column 9) except for the smallest grain sizes. For these magnetically hard grains neglect of the slope of the line CD in Fig.4.5 may lead to a significantly low value of H_{cR} and hence of χ_i and χ by (4.48) and (4.49).

This result provides an explanation for an empirical test for paleomagnetic stability of rocks, suggested by L. G. Parry. We observe that for the grains of highest intrinsic hardness the values in column 9 of Table 4.I exceed the values in column 8, but for all other grains the values are equal. We can therefore make this the requirement for stability:

$$\chi > \left(1 - \frac{H_c}{H_{cR}}\right)/N \tag{4.50}$$

Since for the larger (and presumably less stable) grains $N \approx 4$ and $\chi \approx 0.2$ e.m.u., the condition for stability becomes:

$$H_{cR} < \frac{H_c}{1 - N\chi} \approx 5H_c \tag{4.51}$$

which is Parry's rule.

4.5 DEFLECTION OF MAGNETIZATION IN A STRONGLY MAGNETIC LAYER

For very strongly magnetic rock the integrity of paleomagnetic directions is affected by shape anisotropy of the whole rock body, independently of any intrinsic anisotropy of the rock (Strangway, 1961). A simple calculation suffices to determine the intensity of magnetization at which this effect becomes significant.

We consider the extreme case of an infinite layer of rock (Fig.4.6) which is magnetized by any process which gives magnetization, induced or remanent (T.R.M. or C.R.M.),

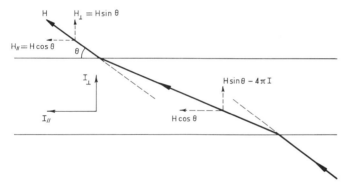

Fig.4.6. A magnetic field is refracted at the boundary of a magnetized or magnetizable medium. The field component parallel to the boundary is the same inside and outside, but the normal field component is reduced by an amount equal to 4π times the normal component of magnetization.

proportional to the inducing field strength. Considering first the problem of induced magnetization, we resolve the external field into components $H_{//}$, parallel to the layer, and $H_\perp$, perpendicular to it. The induced magnetization parallel to the layer is given in terms of the rock susceptibility χ:

$$M_{//} = fI_{//} = \chi H_{//} \tag{4.52}$$

since no self-demagnetizing field operates. However, magnetization perpendicular to the layer, $M_\perp$, produces surfaces of magnetic polarity at the faces and a consequent demagnetizing field $-4\pi M_\perp$, so that:

$$M_\perp = \chi(H_\perp - 4\pi M_\perp) = \frac{\chi H_\perp}{1 + 4\pi\chi} \qquad (4.53)$$

The angle of magnetization with respect to the plane of the layer, ϕ, is therefore related to the angle θ of the external field, H, by:

$$\tan\phi = \frac{M_\perp}{M_{//}} = \frac{1}{1+4\pi\chi} \cdot \frac{H_\perp}{H_{//}} = \frac{\tan\theta}{1+4\pi\chi} \qquad (4.54)$$

If we impose the condition $(\theta - \phi) < 3°$ (at $\theta \simeq 45°$ where $(\theta - \phi)$ is greatest) then we require $(1 + 4\pi\chi) < 1.11$, or $\chi < 9 \cdot 10^{-3}$ e.m.u. This result can now be used in a similar calculation for remanence by noting that this is the limit imposed upon the effective susceptibility of the rock at the time when it acquired its remanence, i.e., at the blocking temperature if the remanence is T.R.M. We can refer to this as the blocking susceptibility, χ_B. The relationship between χ_B and the intensity of remanence of a rock depends upon the mechanism of magnetization, but we may assume that in any rocks so strongly magnetic as to be of concern here ($\chi_B > 10^{-2}$ e.m.u.) the properties are dominated by multidomain magnetite grains, so that we restrict the discussion to T.R.M. and C.R.M. of multidomain materials. Anticipating the discussions in Chapters 7 and 9, we can put $\chi_B = f/N$ for a material containing a volume fraction f of magnetic mineral in grains of demagnetizing factor N. Then, by eq.7.10 and 9.18:

$$M_{TRM} \approx \chi_B H \cdot \frac{I_S}{I_{SB}} \cdot \frac{1}{1+N\chi_i} \qquad (4.55)$$

$$M_{CRM} \approx \chi_B H \cdot \frac{1}{1+N\chi_i} \qquad (4.56)$$

where I_S/I_{SB} is the fractional increase in spontaneous magnetization between the blocking temperature and laboratory temperature, χ_i is the intrinsic susceptibility of the magnetic mineral present and N is the demagnetizing (shape) factor of the grains in which it occurs. Typically $I_S/I_{SB} \simeq 2.5$, $(1 + N\chi_i) \approx 4$ to 6, and if we assume $H = 0.5$ Oe, our calculated susceptibility limit becomes $I_{TRM} < 2 \cdot 10^{-3}$ e.m.u. or $I_{CRM} < 1 \cdot 10^{-3}$ e.m.u. Since C.R.M. may be more important than T.R.M. even in igneous rocks (Chapter 9), we must be prepared to consider the remanence of a rock to be measurably deflected towards the plane of a rock layer (e.g., a sill or dyke) if its remanence exceeds 10^{-3} e.m.u., which is quite commonly the case in basaltic rocks.

Chapter 5

PROPERTIES OF HEMATITE GRAINS

5.1 ANISOTROPY AND COERCIVITY

Fine powders of hematite have very high coercivities, up to 3,000 Oe (e.g. Haigh, 1957; Dunlop, 1970), in contrast with the low values, of order 10 Oe, observed with crystals of both natural and synthetic hematite a few millimetres in size (Sunagawa and Flanders, 1965; Flanders and Remeika, 1965). Since the spontaneous magnetization is restricted to the basal plane, the coercivity of hematite must be due to basal plane anisotropy. Shape anisotropy, even for a needle-shaped grain, could not lead to coercivity exceeding 25 Oe, because the spontaneous magnetization is only 2 e.m.u.; thus the anisotropy must be of crystalline origin. The first attempt to observe it was by Néel and Pauthenet (1952) who compared magnetization curves for different directions of magnetization within the basal plane and concluded that hematite was isotropic within the plane. Ferromagnetic resonance experiments (Anderson et al., 1954; Kumagai et al., 1955; Tasaki and Iida, 1963) detected a triaxial (sixfold) anisotropy in the basal plane, which was consistent with the crystal symmetry but of magnitude corresponding to a coercivity of order 1 Oe and therefore contributed insignificantly to the coercivities of fine powders. However, torque magnetometer measurements on natural single crystals by Banerjee (1963a) found in addition to a small term of the anticipated triaxial form, uniaxial and biaxial components of anisotropy which Stacey (1963) interpreted as arising from the interaction of oriented internal stresses with the fairly strong magnetostriction of hematite ($\sim 8 \cdot 10^{-6}$). The validity of this interpretation was confirmed by Sunagawa and Flanders (1965), who showed that natural single crystals with large internal stresses or tin impurities had strong triaxial anisotropies, and by Porath and Raleigh (1968) who deformed natural single crystals to produce twinning and accompanying strong increases in anisotropy (measured with a torque magnetometer) and coercive force. They attributed the uniaxial and biaxial components of their torque curves to unequal volumes of differently oriented twins. Also Porath (1968) showed that elastic compression in the basal plane produced a strong basal plane anisotropy with uniaxial, biaxial and triaxial components. The uniaxial terms were dominant, as expected for linear compression, and had a magnitude of about 10^3 erg/cm^3 at a stress of 200 kg/cm^2, sufficient to account for a coercivity of 500–1,000 Oe. The crystal deformation in this experiment was entirely elastic, but Porath appealed to twinning, as in the experiments on anelastic deformation, as the explanation of the observed anisotropy, in this case a temporary, "elastic twinning". The stress-induced anisotropy of hematite crystals has been confirmed also by ferromagnetic resonance (Mizushima and Iida, 1966).

The experiments with stressed single crystals of hematite lead directly to the conclusion that its coercivity is due entirely to anisotropy of magnetostrictive origin, resulting from internal stresses (or possibly to stresses resulting from contact with adjacent mineral grains in the case of hematite within a rock). This implies that highly coercive fine powders have high internal stresses and it now appears that, for a reason which is not understood, high internal stresses are an intrinsic property of fine-grained hematite. Quadrupole splitting of Mössbauer spectral lines of chemically prepared fine particles of hematite increases with decreasing grain size (Kündig et al., 1967; Schroeer, 1968) and Yamamoto (1968) reported an increase in lattice parameter with decreasing grain size. Making use also of the pressure dependence of the Morin transition, Schroeer and Nininger (1967) concluded that 50 Å (0.005 μm) grains had an effective internal pressure of minus 180 kbar. Recent measurements by Banerjee (1971) show that the internal stress maintains a stable single-domain state in grains down to 275 Å diameter.

Chevallier and Matthieu (1943) measured hysteresis loops of powdered single-crystal hematite and obtained coercive force as a function of grain size, showing a pronounced maximum at 5–10 μm diameter. However, as Chevallier (1951) recognized, this material appears different from fine-grained hematite produced chemically (e.g., by decomposition of ferric nitrate at 300°C). Apparently the internal stresses within fine grains are different if the grains have been produced by chemical growth from smaller ones than if they are the product of crushed larger grains.

Assuming that the uniaxial magnetostrictive anisotropy in fine-grained hematite has a maximum value of about 10^4 erg/cm^3 (Kündig et al., 1966) for grains in the range of sizes of greatest interest, since the spontaneous magnetization is 2 e.m.u., we expect a maximum coercivity of about $10^4/2 = 5,000$ Oe, which accords well with measurements by Chevallier (1951), Haigh (1957) and Dunlop (1970).

Dunlop (1971) reported measurements of saturation magnetization, I_S, coercivity, H_c, and coercivity of remanence, H_{cR}, of fine-grained hematite prepared by chemical precipitation and subsequently annealed in air at temperatures between 650°C and 950°C. The effect of annealing was to decrease I_S and increase both H_c and H_{cR}. Dunlop interpreted the decrease in I_S to indicate the removal of a moment due to a ferrimagnetic-like spin imbalance resulting from point defects, but very fine grains have enhanced spontaneous magnetization due to increased spin-canting angle relative to bulk material so that sintering of the fine grains could have the same effect. In either case if the anisotropy energy responsible for constraining the spin orientations was unaffected by annealing, H_c and H_{cR} would have increased to maintain the products $H_c I_S$ approximately constant.

5.2 SUSCEPTIBILITY OF HEMATITE

From an extensive study of hematite grains prepared in various ways, Chevallier (1951) concluded that the basal plane spontaneous magnetization 0.4 e.m.u./g or 2 e.m.u./cm^3 was superimposed upon an isotropic susceptibility of $20 \cdot 10^{-6}$ e.m.u. For hematite crystals of low coercivity, the low-field susceptibility within the basal plane is, of course, much larger than this, because a contribution due to alignment of the spontaneous moment dominates the observed susceptibility. However, even after saturation alignment of the spontaneous moment, increasing field causes increasing magnetization by turning the spins nearer to the field direction, i.e., increasing the canting angle and this is the same process as that producing induced magnetization in a field directed along the trigonal axis. We can examine these two processes separately.

First we consider the rotation of spontaneous magnetization by a field H against the restraint of the basal plane anisotropy, which we assume to be uniaxial and to be represented by:

$$E_K = K \sin^2 \phi \tag{5.1}$$

where ϕ is the angle of deflection away from the easy axis and K is anisotropy energy per unit volume. Assuming the field to be applied normally to this axis, but within the basal plane, the field energy of the magnetization (per unit volume of hematite) is:

$$E_H = -H I_S \sin \phi \tag{5.2}$$

I_S being the spontaneous magnetization. The angle ϕ is determined by the minimum energy condition:

$$d(E_K + E_H)/d\phi = 0 \tag{5.3}$$

which gives:

$$\sin \phi = \frac{H I_S}{2K} \tag{5.4}$$

and the susceptibility perpendicular to the easy axis within the basal plane is therefore::

$$\chi = \frac{I_S \sin \phi}{H} = \frac{I_S^2}{2K} \tag{5.5}$$

But in an assembly of randomly oriented crystals the contribution of this process to the average susceptibility is only one third of the result in (5.5), i.e.:

$$\chi = \frac{I_S^2}{6K} \tag{5.6}$$

If $K \approx 10^4$ erg/cm^3, as deduced by Kündig et al. (1966) from ferromagnetic resonance in fine hematite, then we obtain $\chi = 67 \cdot 10^{-6}$ e.m.u., still larger than the susceptibility due to rotation of spins against the exchange forces.

The susceptibility contribution given by eq.5.6 is directly related to coercivity H_c, as in the equivalent calculation for magnetite (section 4.2). Considering the geometrically simple case of a grain with its easy axis parallel to the field, the condition that the field H should suffice to turn the moment against the anisotropy is:

$$H_c > \frac{2K}{I_S} \qquad (5.7)$$

In a random assembly the coercive force is half of this value, i.e.:

$$H_c = \frac{K}{I_S} \qquad (5.8)$$

so that by equations (5.6) and (5.8):

$$\chi H_c = \frac{I_S}{6} = 0.3 \text{ e.m.u.} \qquad (5.9)$$

Referring to the maximum observable coercive force of about 5,000 Oe for very fine grains, mentioned in section 5.1, we obtain for this material $\chi = 60 \cdot 10^{-6}$ e.m.u. which demonstrates that the value of K deduced by Kündig et al. (1966), and used above, is consistent with the 5,000 Oe coercivity.

Apart from its weak ferromagnetism, hematite has an antiferromagnetic contribution to susceptibility, due to the rotation of spins by a field against the exchange forces. This process is considered in section 1.4 and the susceptibility of hematite at low temperatures, i.e., in the perfectly antiferromagnetic state well below the Morin transition, $\sim 2 \cdot 10^{-6}$ e.m.u., is a measure of its contribution at all temperatures below the Curie point. Above the Morin transition there is an additional, stronger contribution to susceptibility arising from the weak ferromagnetism, and therefore restricted to the basal plane. This is inversely related to coercivity by eq.5.9. At the Morin transition itself there is a peak in susceptibility which is consistent with the vanishing of low order anisotropy constants, so that Cinader et al. (1967) explained its value in terms of higher order constants. Above 700°C, the susceptibility of hematite, as quoted by Chevallier (1951) from the work of K. Endo, follows eq.1.48 with $C = 0.085$ e.m.u. and $\theta = 3,727°$.

Very fine grained hematite is superparamagnetic (section 6.3) and Creer (1961) observed this in red sandstones. However, the critical size for the onset of superparamagnetism is very low, because the strong internal stresses of very fine grains noted in section 5.1 increase their magnetic stabilities. These stresses also have the effect of lowering the Morin transition, and in grains smaller than about 250 Å suppressing it altogether. Below about 100 Å the susceptibility is enhanced apparently by a mechanism of superantiferromagnetism, also discussed in section 6.3.

5.3 ANISOTROPY AND ROTATIONAL HYSTERESIS IN HEMATITE-BEARING ROCKS

Hematite commonly occurs as fine flakes or plates whose planes coincide with the crystallographic basal planes. During deposition in water or crystallization under stress there is therefore a tendency of basal planes to align to a common direction, producing a foliated rock. Then, since the spontaneous magnetization is confined to the basal plane of each crystal, the direction of natural remanence in the rock can only coincide with the field direction if the field is parallel to the preferred plane of alignment. Commonly this is horizontal, so that there is an inclination error or deflection of remanence towards the horizontal from any other field direction. Inclination errors have been well studied in laboratory experiments on depositional remanence of sediments (Chapter 8) and occur with titanomagnetite grains as well as hematite. However, the effect can be very much stronger with hematite because of its intrinsic crystallographic anisotropy, whereas the anisotropy of titanomagnetite-bearing rocks depends upon the grain elongations.

From the point of view of anisotropy of susceptibility we can regard the c axis of hematite as magnetically hard and the basal plane easy, but not necessarily equally easy in all basal plane directions because there is in general also a basal plane anisotropy, as discussed in section 5.1. However, the anisotropy of hematite-bearing rocks is usually well represented by an oblate susceptibility ellipsoid, the plane of high susceptibility being the preferred basal plane alignment of the hematite (Howell et al., 1960; Fuller, 1963).

A useful by-product of torque meter measurements of magnetic anisotropy is information about rotational hysteresis, which is the work done due to irreversible magnetization processes when a sample of magnetic material is rotated through 360° in a constant magnetic field. The best method of measurement is to rotate the field about the specimen and to measure the torque curves for both directions of rotation. Then the area enclosed by the two curves is a measure of the irreversible work performed during the two cycles, being proportional to twice the rotational hysteresis loss per cycle, $2W_R$.

The variation of rotational hysteresis loss W_R, with applied field H, has been predicted for an assembly of magnetic particles by various theoretical models. Stoner and Wohlfarth (1948) considered single-domain particles for which the magnetization I_S follows the applied field by a process of coherent rotation. Consider such a process for a prolate ellipsoid (Fig.5.1) in a single-domain state where θ is the angle between H and the easy axis and ϕ between I_S and the easy axis. At low values of H, as θ increases from 0 to 2π, I_S makes only small reversible excursions from the easy axis. For these reversible excursions at low field, H is less than K/I_S, where K is the uniaxial anisotropy constant. For $H \gg 2K/I_S$, I_S follows H without any lag and again the changes in position of I_S are perfectly reversible. Since rotational hysteresis loss, W_R, arises only from irreversible rotations of I_S, it is zero for $H \ll K/I_S$ and $H \gg 2K/I_S$. If H is slowly increased from zero, a critical value H_{crit} is reached at which for a certain critical value of θ (i.e., θ_{crit}

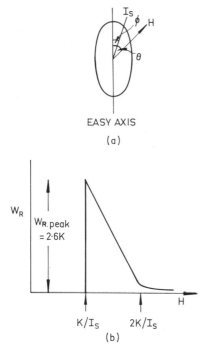

Fig.5.1. Rotational hysteresis for a uniaxial grain represented by a prolate ellipsoid (a) and the dependence of rotational hysteresis loss W_R upon field strength (b).

$< \pi$) ϕ jumps through θ to a value between θ and π. This process constitutes rotational hysteresis, and W_R can be calculated from the initial and final positions of the jump. For an aggregate of aligned single-domain prolate ellipsoids, the maximum value of $W_R = 2.6K$, where K is the uniaxial anisotropy constant. For $H_{crit} < H < 2K/I_S$, W_R slowly decreases from the peak value to zero at $H \gg 2K/I_S$.

In the preceding discussion of coherent rotation it has been assumed that I_S follows H in a linear fashion. In real materials, however, above a certain nucleation field, I_S changes nonlinearly with H, which gives rise to incoherent rotation. In general, incoherent processes lower the value of H_{crit} at which W_R attains the maximum value. When the discussion is extended to include multidomain particles, the situation becomes even more complicated, so that it is not possible to interpret quantitatively the peak value of W_R and H_{crit} for mineral assemblages, but the peak value of W_R is directly proportional to the magnetic fraction in an assemblage and to the anisotropy constants. The value of H at which the peak value of W_R is attained is also directly proportional to the anisotropy field H_A $(= K/I_S)$ of the magnetic fraction. The width of the W_R peak gives an estimate of the spread of anisotropy values in a multicomponent system.

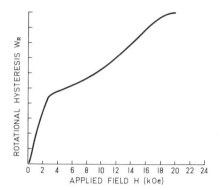

Fig.5.2. Rotational hysteresis as a function of field strength for fine-grained hematite. (After Day et al., 1970.)

In Fig.5.2 the W_R-H curve of a polycrystalline hematite assemblage in plaster-of-Paris is shown. The knee at low field may represent a peak corresponding to irreversible processes due to rotation of spins in the basal plane of individual single crystallites. The peak, which appears to occur at a field greater than 20,000 Oe is consistent with the anisotropy field of hematite as found by Flanders and Schuele (1964) and the independent W_R-H observations of Vlasov et al. (1967b). The increase of W_R with increasing field above 10,000 Oe is typical of hematite and can be used to distinguish it from magnetite or titanomagnetite in natural rocks. Day et al. (1969) have used W_R-H measurements to prove the existence of hematite in highly oxidized basalts. Although optical microscopy could detect the presence of a rhombohedral phase, it was not possible from the optical evidence to say whether it was hematite or ilmenite. Since ilmenite is paramagnetic at room temperature, the rhombohedral phase must be hematite. Furthermore, the observed high stability of N.R.M. in such rocks could be at least partly due to hematite, and not merely due to subdivided magnetite, as postulated by Strangway et al. (1968).

5.4 TITANOHEMATITES

Titanohematites ($Fe_{2-x}Ti_xO_3$) have been studied by the Tokyo and Sydney groups (Uyeda, 1958; Nagata und Akimoto, 1965; Westcott-Lewis and Parry, 1971a, b; Westcott-Lewis, 1971) with particular attention to the occurrence of self-reversed TRM for compositions $x \approx 5$. Both coercive force and spontaneous magnetization show interesting variations with the composition parameter, x, temperature, T, and grain size.

Between the values of $x = 0$ and $x = 0.5$, the crystal structure is $R\bar{3}C$, as in pure hematite. The addition of titanium causes the reduction of a fraction of the ferric ions to the ferrous state, since $x(Fe^{2+}Ti^{4+})$ replaces $2xFe^{3+}$. The ferrous ions impart large magnetocrystalline anisotropy to the compounds between $x = 0$ and $x = 0.5$ causing an initial increase in coercivity with x, although increasing spontaneous magnetization with x then leads to a decrease in coercivity as $x \to 0.5$ (Fig.2.7). However, other factors complicate the properties of titanohematites, especially the exsolution of a Ti-poor phase (hematite) and a Ti-rich phase (ilmenite) as mentioned in section 2.6. Stresses occur at the boundaries of the exsolution lamellae, which, although partly relieved by dislocations (Shive and Butler, 1969), cause an anisotropy of magnetostrictive origin which enhances the coercivity; thus coercivity is dependent upon the thermal history of a sample. In materials such as the Allard Lake titanohematites, studied by Carmichael (1959), individual exsolution lamellae are visible under a microscope. A further problem is the ordering of Ti^{4+} ions on the lattice sites, causing a variation of spontaneous magnetization with order (also referred to in Chapters 2, 12). Spontaneous magnetization remains small, as in hematite, for $x < 0.5$, but increases very strongly for $x > 0.5$ to a maximum of about 150 e.m.u. cm^{-3} at $x = 0.66$ (Westcott-Lewis and Parry, 1971a) due to the onset of normal ferrimagnetism, rather than the parasitic ferromagnetism observed in hematite.

In the size range 5 μm $< d <$ 30 μm, Westcott-Lewis and Parry (1971a, b) found ilmenite-hematite grains to have coercivities varying with grain size, d, as d^{-n} with $n \approx 0.4$, as in magnetites, and remanence showed characteristically pseudo-single domain type properties with TRM varying as d^{-1}. Reversed thermoremanence was observed only with the composition range $0.51 < x < 0.73$ (Westcott-Lewis and Parry, 1971b) and only with grains larger than 5 μm; smaller grains gave normal thermoremanence in all cases. Westcott-Lewis (1971) suggested that self-reversal was due to the exsolution of ordered and disordered volumes of order 5 μm in size and was therefore inhibited in smaller grains.

5.5 THE MAGHEMITE-TO-HEMATITE TRANSITION

In section 2.3 we have discussed the low-temperature ($T \sim 250°C$) origin of maghemite (γFe_2O_3) by slow oxidation of magnetite (Fe_3O_4). We also emphasized that maghemite is a metastable compound. At temperatures greater than 350°C (the actual temperature depends on the grain size, impurities and the ambient atmosphere), maghemite irreversibly converts to hematite. Magnetically, the transition is marked by a dramatic fall in the spontaneous magnetization, I_S, from a value of about 350 e.m.u. for maghemite to 2 e.m.u. for hematite. Polycrystalline hematite, produced by the transformation of maghemite, has been found in highly oxidized continental lavas, ocean sediments and in dredged submarine basalts. The mechanism of the maghemite-to-hematite transition, and its dependence upon pressure and impurity content are important for two reasons. Firstly, a confirmation that the hematite is derived from maghemite leads to the conclusion that the N.R.M. of such hematite is a C.R.M., and the age of the magnetization is open to question. Secondly, the observed transition in a particular rock may lead to a reconstruction of its history of pressure, temperature and other ambient conditions.

The dependence of the transition temperature on impurities was studied by Bénard (1939), who prepared a series of substituted maghemite compounds with increasing amounts of magnesium, manganese and aluminium. He first prepared the corresponding substituted magnetites ($Fe_{3-x} Me_x O_4$, Me = impurity element) and subjected them to low-temperature oxidation. All of these substitutions raised the maghemite-to-hematite crystallographic transition temperature and lowered the magnetic Curie temperature. Michel and Chaudron (1935) had previously used the same technique to prepare sodium-doped maghemites, whose transition temperatures were much higher than the corresponding Curie points. They then extrapolated the Curie points to zero sodium content, obtaining an extrapolated Curie point of 675°C for pure maghemite, this being unobservable because it is much higher than the maghemite → hematite transition. De Boer and Selwood (1954) studied the variation in transition temperature with the substitution of trivalent aluminium, gallium and lanthanum. By measuring the rate of decrease in magnetization with time for constant temperatures, they arrived at different values of the activation energy of transition for the three different impurity ions and found that aluminium increases it, lanthanum decreases it and gallium left it unaffected. Banerjee (1963b) pointed out that gallium preferentially occupies the tetrahedral sites in a spinel, replacing the iron atoms and so causing an increase in saturation magnetization, as observed by De Boer and Selwood (1954). Since it does not occupy vacancies it cannot affect the transition. On the other hand, the introduction of aluminium does not affect the spontaneous magnetization, showing that it occupies vacancies in the maghemite structure, making the transition to hematite more difficult. Lanthanum ions decreased the spontaneous magnetization indicating that they displaced octahedral Fe^{3+} ions, but La^{3+} are large ions and distort the lattice, making the transition to hematite easier.

However, although not all impurities increase the stability of maghemite, it is evidently normal for impure maghemite to be much more stable than the pure mineral. Thus Wilson (1961) found a highly stable maghemite in the weathered lateritic lavas of Northern Ireland, attributing it to the presence of aluminium.

A surprising claim was made by Aharoni et al. (1962) regarding the stability of maghemite in helium gas at elevated temperatures. They reported that while heating in air produced the usual transition to hematite, heating in helium left about 2/3 of the original maghemite unchanged. This led them to postulate that ordinary maghemite is a mechanical mixture of two different chemical compounds: (A) $Fe_{5/3} \square_{1/3} O_4$ and (B) $H_{0.5} Fe_{2.5} O_4$, the proportion of $B : A$ being 2 : 1. They claimed that the hydrogen ferrite, $H_{0.5} Fe_{2.5} O_4$, is stable in helium and hence does not transform to hematite on heating. Further, they analyzed for hydrogen and found the amount predicted from magnetic measurements. If the hydrogen ferrite model is correct, the ratio $B : A = 2 : 1$ is only approximate because the spontaneous magnetization of $H_{0.5} Fe_{2.5} O_4$ cannot be the same as that of $Fe_{5/3} \square_{1/3} O_4$. At absolute zero, ($I_S$ of B)/(I_S of A) = 3/4. If this ratio is assumed also for room temperature, Aharoni et al.'s magnetic data should lead to a mechanical mixture of the ratio of about 2.2 : 1, but their chemical analysis of hydrogen is not compatible with this value, so the suggestion is subject to obvious doubt. The effect of the ambient gas is considered further below.

The effect of pressure seems more straightforward. An increment of pressure hastens the transition because of its effect on vacancies. Kushiro (1960) has obtained a hydrostatic pressure dependence of 3.5° per kbar, suggesting that burial and subsequent heating of sediments containing maghemite would hasten the probability of hematite formation. Such hematite will carry a C.R.M. due to volume growth in the presence of the earth's magnetic field. Banerjee (1969) attempted to combine the available pressure dependence data and his field dependence data to describe the nature of the maghemite-to-hematite transition thermodynamically. Preliminary differential thermal analysis (D.T.A.) data for a particular sample gave a transition at 548°C in flowing nitrogen gas while a magnetic extrapolation experiment suggested a Curie point only 6° lower, at 542°C, indicating a near coincidence of the transition point and the Curie point and allowing the possibility that the transition was of first order. The equivalent of the Clausius-Clapeyron equation for magnetic first order transitions is:

$$\frac{\Delta \theta_c}{\Delta H} = \frac{\theta_c}{L} \Delta M \qquad (5.10)$$

where θ_c = Curie point. H is the field strength, L = latent heat and ΔM = incremental decrease in magnetization at the Curie point. The field dependence experiment led to the value L = 0.18 cal/g. A second value of L was obtained from the pressure-dependence data of Kushiro (Fig.5.3), using the usual Clausius-Clapeyron equation:

$$\frac{\Delta \theta_c}{\Delta P} = -\frac{\theta_c}{L} \frac{\Delta V}{V} \qquad (5.11)$$

the transition temperature being again written as θ_c and $\Delta V/V$ denotes the fractional change in volume at the transition. This gives L = 0.08 cal./g. But the exothermic D.T.A. peak leads to L = 22 cal./g, which is completely incompatible with the magnetic data. The disagreement shows that the transition cannot be a first-order one; but the abrupt fall of I_s near the Curie point precludes a second-order transition. Thus we have to conclude that it is of a monotropic nature, occurring slowly at all values of temperature. Because the free energy G_γ of maghemite is greater than G_α of hematite, the transition can take place even at room temperature, if the grain size, impurity content and ambient gas are appropriate. ...

Table 5.I gives D.T.A. peaks for two 1-μm size maghemite samples in different ambient gas atmospheres. The hydrogen ferrite model of Aharoni et al. cannot explain the shift in transition temperature with ambient gas as represented in Table 5.I. Both of the samples in the table were found to be most stable in oxygen, while according to Aharoni et al., the result should be the opposite.

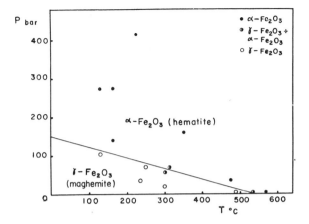

Fig.5.3. Pressure dependence of the maghemite–hematite transition temperature. (Reproduced by permission, from Kushiro, 1960.)

TABLE 5.I

Variation of DTA peak with ambient gas atmosphere for maghemite samples

Sample	Nitrogen	Static Air	Oxygen
EMI-acicular	548	589	605
Mapico	457	482	501

The numerous effects and impurities which influence the maghemite → hematite transition so complicate the problem that it is difficult to present any quantitative explanation. However, when maghemite and hematite are found to coexist in nature, the conclusion is that the hematite *is* a transformation product of the maghemite, even though the maghemite may contain "stabilizing" impurities or might have had a non-oxidizing environment in the past, as under certain oceanic conditions.

Chapter 6

THERMAL ACTIVATION EFFECTS

6.1 THE CONCEPT OF BLOCKING TEMPERATURE

The basic equation of thermal activation phenomena gives the probability $dP\,(\ll 1)$ of a change occurring in a short time dt in terms of the activation energy, E, that is the "height" of the energy barrier which must be overcome thermally for the change to occur, and the absolute temperature T:

$$dP = C \exp(-E/kT)\, dt \qquad (6.1)$$

where k is Boltzmann's constant and C is a numerically large frequency factor. Estimates of C have been in the range 10^9 to 10^{13}/sec, but for the magnetic processes which are of interest here we now favour the value $C = 10^8$/sec, which is approximately the frequency of a spin wave of half wavelength equal to a domain-wall thickness in magnetite. The precise value is not important, but the significant fact is that it corresponds to a time interval very much shorter than any observations in rock magnetism. We can see immediately that for any process which is not effectively instantaneous the barrier energy must substantially exceed kT. Then dP/dt is very sensitive to small variations in E/kT, as was emphasized by Néel (1955). The characteristic time for a process, which appears as a relaxation time in theories of magnetic viscosity and thermoremanence in single domains, is:

$$\tau = \left(\frac{dP}{dt}\right)^{-1} = \frac{1}{C}\exp\left(\frac{E}{kT}\right) \qquad (6.2)$$

Taking C^{-1} as 10^{-8} sec we find that, $\tau = 1$ sec for $E/kT = 18.4$, and $\tau = 10^{17}$ sec $(3 \cdot 10^9$ years – the age of the oldest known rocks), for $E/kT = 57.6$. Thus a relatively small change in temperature (or barrier energy E) has a dramatic effect on relaxation time. This leads immediately to Néel's concept of a *blocking temperature*, which is central to all theories of thermoremanent magnetization.

Since the characteristic time for thermal activation is such a strong function of temperature by eq.6.2, it follows that as a magnetic body cools so it passes through a narrow temperature range over which there is a dramatic change in the rate of thermal activation. Thus at a particular temperature single domains may undergo frequent spontaneous (thermally excited) reversals of magnetization but at a temperature only slightly lower the average time for reversal becomes much longer than the time scale of observation. They thus change quite suddenly from the superparamagnetic state to a stable, permanently magnetized state. The temperature at which this occurs, assumed to be sharp, is referred to as the blocking temperature. Although this explanation implies a very simplified model, the existence of definite blocking temperatures for different materials is well-established in numerous experiments on partial thermoremanent magnetizations (section 7.6). It is therefore worth considering the definition of blocking temperature more carefully, in particular to examine the suggestion (Stacey 1963, p.129) that the effective blocking temperature is a function of cooling rate, being lower for slow cooling, since blocking temperature is really the temperature at which the relaxation time for thermal activation becomes longer than the time scale of cooling.

We can consider the problem quantitatively by considering a material to be cooling at a constant rate through its blocking temperature, so that as a function of time t its temperature is:

$$T = T_B(1 - t/\tau'), \text{ i.e.,} \tag{6.3}$$

$$\frac{dT}{dt} = -\frac{T_B}{\tau'}$$

τ' being the characteristic cooling time and T_B is the blocking temperature. This fixes the zero of the time scale as the instant at which $T = T_B$ (but with T_B still to be precisely defined) and τ' is the time required to reach $0°K$ if the cooling continued at the same rate. We can now substitute for T by eq.6.3 in eq.6.1 and then define T_B as the value of T for which the integral of dP during cooling below T_B becomes unity:

$$\int dP = C \int_0^{\tau'} \exp\{-E/kT_B(1 - t/\tau)\} dt = 1 \tag{6.4}$$

Exact evaluation of this integral can be obtained only in terms of an inconvenient series, but as we are concerned with integration over the range $E/kT > 20$, a considerable simplification becomes valid[1] and we obtain the result:

$$C\tau' \frac{kT_B}{E} \exp\left(-\frac{E}{kT_B}\right) = 1 \tag{6.5}$$

Thus we can define blocking temperature, at any particular cooling rate represented by τ', as the value of T_B which satisfies eq.6.5. The dependence of T_B upon τ' by eq.6.5 may be significant.

Thus, although we think of blocking temperatures in terms of a simple model of thermal activation, as represented by eq.6.1 or 6.2, the actual phenomena are more complex than any of our theories and we can do little better than treat blocking temperature in a phenomenological way as the temperatures at which domains or domain walls become blocked or prevented from making spontaneous changes.

---

[1] $\int_0^{x_0} \exp\left(-\frac{1}{x}\right) dx \approx x_0^2 \exp\left(-\frac{1}{x_0}\right)$ for $x_0 \ll 1$.

6.2 MAGNETIC VISCOSITY AND PALEOMAGNETIC STABILITY

The term "magnetic viscosity" antedated modern theories of magnetism and appealed to a hypothetical fluid-like drag on the rotations of elementary (atomic) magnets, causing a lag in their response to an applied field. Three distinct causes of magnetic viscosity are now known. One is concerned with eddy currents which oppose magnetic flux changes in metals and has no relevance to rock magnetism. A second is associated with ionic reordering of a magnetic lattice. This occurs in carbon steel, in which carbon atoms migrate between alternative interstitial sites of the iron lattice, choosing preferentially the sites favoured by the magnetostriction of the lattice but lagging in their response to a change in magnetization. This process is related to the phenomena of chemical remanence (Chapter 9) and of self-reversals (Chapter 12) in ilmenite-hematite solid solutions. The most important of the viscous effects in rock magnetism arises from thermal activation of domains.

The response of an assembly of independent, identical single domains to a change in field follows directly from the probability equation (6.1). The simplest case is the randomization of moments of an initially aligned assembly of uniaxial grains after removal of the aligning field. If we suppose that at a time t after removal of the field there are n moments still parallel to the original direction and n' anti-parallel then the fractional magnetization of the assembly is:

$$\frac{M}{M_0} = \frac{n-n'}{n+n'} \tag{6.6}$$

and since $(n + n')$ is constant $dn'/dt = - dn/dt$, so that:

$$\frac{d}{dt}\left(\frac{M}{M_0}\right) = \frac{2}{n+n'}\frac{dn}{dt} \tag{6.7}$$

But n is diminished by reversal of the parallel domains and increased by subsequent reversals of the n' reversed domains, both processes occurring at a rate given by eq.6.1:

i.e., $$\frac{dn}{dt} = -(n-n')C \exp(-E/kT) \tag{6.8}$$

Thus:

$$\frac{d}{dt}\left(\frac{M}{M_0}\right) = -\frac{M}{M_0}\{2C \exp(-E/kT)\} \tag{6.9}$$

which integrates to:

$$M = M_0 \exp(-t/\tau) \tag{6.10}$$

where:

$$\tau = \frac{1}{2C} \exp(E/kT) \tag{6.11}$$

The time constant τ for restoration of equilibrium of the assembly is changed somewhat if a small field is applied, shifting the equilibrium magnetization to non-zero M/M_0, but the approach to equilibrium is still exponential with time.

No clear example of exponential decay of magnetization has been reported although Banerjee (1971) suggested that thermally activated diffusion of Fe^{2+} ions in titanomagnetites may explain the sharp decrease of magnetization within the central magnetic anomaly over the Reykjanes ridge. As with measurements on laboratory materials, observations of magnetic viscosity in rocks have almost invariably yielded a log (time) relationship. This is technically possible in an assembly of independent single domains if they have an appropriate spectrum of barrier energies, but it must be regarded as a completely implausible explanation for the generality of the log (time) law. Rather we must suppose that the properties of single domains occurring in nature are strongly affected by grain interactions, a conclusion confirmed independently by Dunlop (1968). They thus behave in some respects like multidomains, for which the log (time) law of magnetic viscosity was derived by Stacey (1963). Stacey's theory is followed here, but we note that Tropin and Vlasov (1966) claim that impurity diffusion may also be significant.

Magnetic viscosity in multidomains is due to thermally activated domain wall movements. A simple analytical treatment of the problem considers a domain wall to be moving through a sequence of sinusoidal potential wells which, with a biassing field, are represented by eq.4.26, so that for small biassing fields the potential barriers for movement of the wall in opposite directions are approximately:

$$E, E' = E_0 \mp AHI_S t' \tag{6.12}$$

where the change in magnetic moment due to one Barkhausen jump of the wall is:

$$\Delta m = 2AI_S t', \tag{6.13}$$

being the reversal of magnetization I_S in a volume At'. The rate of change of the total moment in a body with n similar domain walls is the difference between the rates of change due to Barkhausen jumps in the two opposite directions (aided by the field H and opposed by it) multiplied by the magnitude of each increment:

$$\begin{aligned}\frac{dm}{dt} &= n\Delta m \left[C \exp(-E/kT) - C \exp(-E'/kT) \right] \\ &= 2n\Delta m\, C \exp(-E_0/kT) \sinh\left(\frac{H\Delta m}{2kT}\right)\end{aligned} \tag{6.14}$$

Since we are considering a multidomain grain, H and m are related. The field H considered here is the internal field H_i acting on the domain walls and is given by eq.4.14 and $m = VI$ for a grain of volume V, so that:

$$H = H_i = H_e - NI = H_e - \frac{N}{V}m \tag{6.15}$$

Eq.6.14 shows that the grain tends towards the equilibrium magnetization $I = H_e/N$, at which $H_i = 0$. m increases if H_i is positive and decreases if H_i is negative. m (or I) changes logarithmically with time if:

$$|H_i \Delta m/kT| \gg 1 \tag{6.16}$$

i.e., the field-dependent part of the activation energy substantially exceeds kT. To preserve this condition during integration of eq.6.14 we integrate with respect to H_i, rather than m, by writing:

$$\frac{dm}{dt} = -\frac{V}{N} \cdot \frac{dH_i}{dt} \tag{6.17}$$

Since H_i can have either sign, the hyperbolic sine term cannot be approximated by a single exponential term and we integrate (6.14) to obtain:

$$\ln \left\{ \frac{\left|\tanh\left(\frac{\Delta m H_i}{4kT}\right)\right|}{\left|\tanh\left(\frac{\Delta m H_{io}}{4kT}\right)\right|} \right\} = -\frac{n}{V} \cdot \frac{N(\Delta m)^2}{kT} C \exp(-E_0/kT) \cdot t \tag{6.18}$$

which, for the condition (6.16) reduces to:

$$|H| = |H_0| - \alpha \ln(1 + \beta t) \tag{6.19}$$

with:

$$\alpha = 2kT/\Delta m$$

and:

$$\beta = \frac{n}{V} \frac{(\Delta m)^2 N}{2kT} C \exp\left(-\frac{E_0 + \Delta m \cdot H/2}{kT}\right)$$

Condition (6.16) implies $\beta t \gg 1$ for all conveniently usable experimental times, so that:

$$\frac{d|H_i|}{d(\ln t)} = -\alpha \tag{6.20}$$

and therefore:

$$\frac{d}{d(\ln t)}\left(\left|I - \frac{H_e}{N}\right|\right) = -\frac{\alpha}{N} = -\frac{2kT}{\Delta m N} = -S \tag{6.21}$$

S has been referred to as the viscosity coefficient of the material, especially by Néel (1955). Eq.6.21 is in accord with observations which show that over a wide range of experimental parameters magnetization varies linearly with the logarithm of time and that the rate of change is proportional to absolute temperature. The most convincing demonstration of the validity of these conclusions is by Shimizu (1960), whose data are reproduced in Fig.6.1.

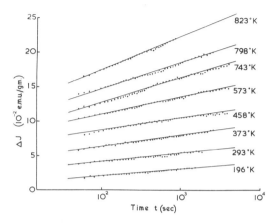

Fig.6.1. Shimizu's (1969) measurements of progressive viscous magnetization of 2 μm magnetite grains due to the application of a 3.3 Oe field. The zero of the ordinate scale is arbitrary, relative displacements of the graphs being arranged for convenience in the drawing.

The nature of the condition (6.16) which must be satisfied if eq.6.19 to 6.21 are to be valid, requires some emphasis. Magnetic viscosity is most readily observed in the decay of saturation remanence or after the sudden application of a field comparable to the coercive force of a material. Then $H_i \Delta m \approx E_0$, the intrinsic barrier energy for the domains or domain walls and since $E_0/kT > 20$ in order that magnetic changes be not more rapid than laboratory measurements could observe, condition 6.16 is clearly satisfied in these experimental situations. However, the internal fields associated with natural remanence are normally so much smaller that (6.16) does not apply and any decay of natural remanence is not logarithmic with time. Nevertheless we expect a specimen in which viscous effects are hard to observe to be magnetically stable and we can make a quantitative extrapolation from measurements of magnetic viscosity at saturation remanence to deduce the stability of natural remanence in simple cases.

We can adopt as a criterion for paleomagnetic stability a result obtained in section 6.1:

$$\frac{E_0}{kT} > 40 \tag{6.22}$$

and since eq.4.27 can be written:

$$E_0 = \frac{1}{\pi} H_c \Delta m \tag{6.23}$$

and $(\Delta m/kT)$ is given in terms of the magnetic viscosity coefficient S by (6.21), it follows that (6.22) can be written:

$$\frac{2}{\pi} \frac{H_c}{NS} > 40 \tag{6.24}$$

Since S is most conveniently measured at the point of saturation remanence, I_{RS}, it is more convenient to express H_c in terms of I_{RS} by eq.4.44, so that:

$$\frac{2}{\pi} \frac{I_{RS}}{S} > 40$$

or:

$$\frac{S}{I_{RS}} < \frac{1}{20\pi} \tag{6.25}$$

This is a very simple stability condition to apply because it requires only the measurement of saturation remanence and its decay with time. Putting $I = I_{RS}$ and $H_e = 0$ in (6.21), we reduce the stability condition to:

$$-\frac{1}{I_{RS}} \frac{dI_{RS}}{d(\ln t)} = -\frac{d(\ln I_{RS})}{d(\ln t)} = -\frac{d(\log_{10} I_{RS})}{d(\log_{10} t)} < 0.0159 \tag{6.26}$$

The absolute value of I_{RS} is not required; thus it is sufficient to use for this purpose the saturation remanence of a whole specimen, without regard for the concentration of magnetic mineral.

The condition 6.26 is derived for an assembly of similar grains and has a direct application only to simple rocks with single magnetic constituents. It demonstrates the relationship between viscosity measurements and paleomagnetic stability for individual constituents of a rock.

6.3 SUPERPARAMAGNETISM AND SUPERANTIFERROMAGNETISM

A single domain grain above its blocking temperature has a magnetic moment μ, which undergoes repeated thermal reorientation, but in the presence of a field H has an average alignment determined by the Boltzmann parameter $(\mu H/kT)$, as for a classical (Langevin) paramagnetic material. Since it has no stable remanence but effectively adopts the equilibrium average magnetization corresponding to the ambient value of H/T, its behaviour is paramagnetic in nature. But since μ is much larger than the atomic moments of a normal paramagnetic material the alignment in a field is much stronger, i.e., the susceptibility may be much higher. Hence the use of the term superparamagnetism to describe its properties. Sufficiently small grains may be superparamagnetic at room temperature and below and the phenomenon has been recognized in both laboratory materials (Bean and Livingston, 1959) and rocks (Creer, 1961).

For each of a number of types of grain the size below which they are superparamagnetic at any temperature may be calculated by assuming some suitable short time (say 1 sec) for τ in eq.6.2, so that:

$$\frac{E}{kT} = 18.4 \qquad (6.27)$$

For a spherical grain of magnetite of volume V the energy barrier opposing changes in domain direction is:

$$E = \left(-\frac{K_1}{12} - \frac{K_2}{27}\right) V = 1.3 \cdot 10^4 \, V \text{ erg} \qquad (6.28)$$

so that at 300°K ($kT = 4.14 \cdot 10^{-14}$ erg) we obtain a critical volume $V = 5.8_6 \cdot 10^{-17}$ cm³, corresponding to a diameter $d = 4.8_6 \cdot 10^{-6}$ cm (0.05 μm). Dunlop (1973a) reported an experimental estimate of 0.03–0.035 μm for the threshold size for superparamagnetism of magnetite grains at room temperature from the saturation remanence at 77°K and 300°K of a sample with a known distribution of fine grain sizes. However, this probably overestimates the threshold size for truly spherical grains, because the measured grains appeared to have some shape anisotropy. Thus there may not be a significant stable single domain size range for spherical grains of magnetite.

However, for elongated single domains the situation is different. Their properties are dominated by shape anisotropy, which is much stronger than the crystalline anisotropy, and so depresses the critical size for superparamagnetism, and at the same time the critical size for the single domain structure is increased by the reduction in magnetostatic energy for an elongated grain magnetized along a long axis. Evans and McElhinny (1969) have thoroughly studied the very fine elongated grains of magnetite occurring as inclusions in the pyroxenes of a South African gabbro and shown them to behave as stable single domains with coercivities dominated by shape anisotropy (section 4.1). For a prolate grain of volume V, magnetic moment $\mu = V I_S$ and demagnetizing factors N_a, N_b N_b the energy barrier is:

$$E = \tfrac{1}{2} (N_b - N_a) I_S^2 \, V \qquad (6.29)$$

In the limiting case of a very elongated grain $(N_b - N_a) \rightarrow 2\pi$ and for magnetite ($I_S = 480$ e.m.u.) this leads to a super-paramagnetic critical volume $V = 1.05 \cdot 10^{-18}$ cm³, smaller by a factor 50 than the critical volume for a sphere (assuming magnetization reversal to be coherent, but rather less if domain wall nucleation and propagation occur).

The contribution of superparamagnetic grains to the susceptibility of a material may be illustrated by considering it to contain a volume fraction $f \ll 1$ of grains of magnetite of individual volumes $V = 10^{-17}$ cm³ and randomly oriented axes. The grain moments are $\mu = VI_S$. At a temperature T and in a field H, such that $\mu H/kT \ll 1$, i.e., well removed from paramagnetic saturation, the fractional alignment is:

$$\frac{M}{M_S} = \frac{\mu H}{3kT} \tag{6.30}$$

(≈ 0.015 at room temperature), as for a classical paramagnet by Langevin's theory (Bean and Livingston, 1959). M is the magnetization per unit volume and is related to the saturation value M_S where:

$$M_S = fI_S \tag{6.31}$$

The resulting susceptibility is thus:

$$\chi = \frac{M}{H} = \frac{fvI_S^2}{3kT} \approx 18f \text{ e.m.u.} \qquad (f \ll 1) \tag{6.32}$$

which for this particularly favourable case is very much greater than the susceptibility of a material with the same concentration of normal ferromagnetic grains. Thus, a small concentration of superparamagnetic particles can affect the susceptibility quite remarkably. However, simple superparamagnetism of non-interacting single domains is not normally observed in rocks. The properties are modified by grain interactions, which influence the estimates of grain size from the temperature and field dependences of susceptibility, as in the measurements of Creer (1961).

Very fine grained antiferromagnetics have enhanced susceptibilities of superparamagnetic type, a phenomenon known as *superantiferromagnetism*. Néel (1962) explained it as a result of net magnetic moments of fine grains due to statistical inequality of the numbers of cations on the two opposite sub-lattices of the antiferromagnetic structure. Although its existence has been demonstrated experimentally (Cohen et al., 1962) it is not without puzzling features. The simple idea of a regular shaped grain with unequal numbers of layers of the two opposite lattices can hardly be valid; it is possible that very high stresses as observed in hematite (sections 5.1 and 5.2) occur quite generally in fine-grained antiferromagnetics and deflect atomic moments sufficiently to increase the canting angle and hence the spontaneous magnetization of very fine grains.

Chapter 7

THERMOREMANENT MAGNETIZATION (TRM)

7.1 INTRODUCTION

The property of thermoremanence is the most important single magnetic property of rocks. Without it this book would never have been written and its authors probably not be engaged upon research in rock magnetism. The combined facts that igneous rocks are magnetized by the earth's field as they cool and that the TRM which they thus acquire is frequently very stable and allows us to deduce the direction (and magnitude) of the field during cooling have led to a major revolution in geophysical thinking. Certainly chemical remanence (Chapter 9) is also an important basis of paleomagnetism, but it is much more difficult to reproduce in the laboratory, is less direct and obvious and is generally less well understood. Thus the results of measurements on sediments of more than limited age would perhaps still be seriously disputed if it were not for their coincidence with data from igneous rocks. It is the directness and essential simplicity of the phenomenon of thermoremanence which established a physical basis for paleomagnetism.

In all of the present discussion of thermoremanence we are concerned with TRM induced in fields which are small compared with the room temperature coercivities of minerals. High field thermoremanence poses additional problems and is not yet the subject of a satisfactory theory, although its ultimate saturation value must be the same as the saturation remanence induced in any other way (section *4.4*). However, since we are concerned primarily with TRM induced by fields comparable with that of the earth, i.e., not more than 1 Oe, the low field condition is the one of particular interest. This allows an important simplification because in terms of domain theory (Chapter 3) the external field energy of a grain is small compared with the internal contributions to magnetic energy and therefore imposes only a minor perturbation on domain structure. All of the conclusions about domain structure and magnetic properties remain valid and we merely introduce a slight bias of magnetization; this is a statistical alignment of single domains or small displacements of domain walls in multidomains.

7.2 NÉEL'S (1955) THEORY OF TRM IN SINGLE DOMAIN GRAINS

The theory of thermoremanence in independent single-domain grains follows directly from the discussion of superparamagnetism (section *6.3*). The TRM induced in an assembly of single domains is determined by the superparamagnetic susceptibility at the blocking temperature T_B, because it is the magnetization at T_B which is "frozen in" by further cooling. The theory of TRM in single-domain grains is central to the subject, quite apart from its historical contribution. This is partly because it is easier to follow than multidomain theories but also because it is basic to the theory of pseudo-single domain effects (section 7.3), which are particularly important in paleomagnetism, as well as applying directly to the rarer stable single domains.

We restrict our interest to uniaxial single domains. As shown in Chapter 6 it is only elongated grains of magnetite, whose coercivities are dominated by uniaxial shape anisotropy, which are large enough to be magnetically stable (i.e., not superparamagnetic) at ordinary temperatures. In hematite grains the uniaxial character is due to stress fields which magnetostrictively select easy directions of magnetization within the basal planes. Thus we consider initially an assembly of uniaxial single-domain grains with moments $\mu_B = VI_{SB}$ at the blocking temperature T_B, and with easy axes all aligned at an angle θ to a small applied field H. At any instant their moments will be strongly concentrated about the two opposite easy directions, but with a distribution of moments in all directions, as they are still able just to cross the potential barriers separating the easy directions. As an approximation which we re-examine later, let us suppose that the moments are entirely confined to the two easy directions; then the proportions with the two orientations are given directly by the difference between the two energies. The orientation nearer to the field has the lower energy and the difference is.

$$\Delta E = 2\,\mu_B\,H\cos\theta \tag{7.1}$$

Thus the ratio of the number in the further direction, n_-, to the number in the nearer direction, n_+, is:

$$\frac{n_-}{n_+} = \exp[-2\mu_B\,H\cos\theta/kT_B] \tag{7.2}$$

and the total number (fixed) is:

$$n = n_+ + n_- \tag{7.3}$$

Thus the fractional net magnetization in the + direction is:

$$\frac{M}{M_S} = \frac{n_+ - n_-}{n} = \frac{1 - \exp[-2\mu_B\,H\cos\theta/kT_B]}{1 + \exp[-2\mu_B\,H\cos\theta/kT_B]} = \tanh(\mu_B H\cos\theta/kT_B) \tag{7.4}$$

On cooling below T_B all of the orientations of the moments remain fixed as their spontaneous magnetization increases so that M/M_S gives the TRM in the θ direction at room temperature. Since we are normally concerned with assemblies of randomly oriented grains we must integrate (7.4) over all θ (0 to $\pi/2$):

$$\frac{M_{TRM}}{M_S} = \int_0^{\pi/2} \tanh(\mu_B\,H\cos\theta/kT_B)\cos\theta\sin\theta\,d\theta \tag{7.5}$$

The series solution appropriate to small values of $(\mu_B H/kT_B)$ is:

$$\frac{M_{TRM}}{M_S} = \frac{\mu_B H}{3kT_B}\left[1 - \frac{1}{5}\left(\frac{\mu_B H}{kT_B}\right)^2 + \frac{2}{35}\left(\frac{\mu_B H}{kT_B}\right)^4 - \frac{17}{945}\left(\frac{\mu_B H}{kT_B}\right)^6 + ...\right] \quad (7.6)$$

Normally only the first term is of interest; the subsequent terms represent the tendency to saturation. The general form of eq.7.5 is seen in the tabulation of this function in Appendix 1.

We can now reconsider the approximation that the moment of each grain was restricted to its two easy directions. Abandoning this assumption the assembly of randomly oriented grains is simply a superparamagnetic material, whose susceptibility in low fields is given by (6.30) which at T_B corresponds precisely to the first term of (7.6). As the temperature is lowered the distributions of the moments in the two opposite hemispheres for each grain-orientation cluster more closely about the two easy directions and are no longer able to cross the barrier between easy directions, that is the equatorial plane is completely unpopulated. But even at the blocking temperature the number of moments instantaneously within 45° of the equatorial plane is (from the Boltzmann distribution) only about 1 in 10^4 (the barrier energy being about $20\,kT_B$). The approximation made in deriving eq.7.6 is therefore an extremely good one.

The limitations of this theory are two-fold. Firstly, its applicability is to a physical situation which is probably rare. It cannot account for the TRM of interacting single domains because the interactions tend to produce mutual cancellation of the alignments. Secondly, even for a material to which it could apply, a quantitative check is very difficult to make because it requires grains in a very limited size range and with precisely constant shape in order that both μ_B and T_B should be the same for all of them. This is a requirement which is virtually impossible to meet. Nevertheless the theory does have a quantitative application to pseudo-single domain TRM. An improved approach to this problem, appealing to grain surface effects, is presented in section *7.4*.

7.3 TRM IN LARGE MULTIDOMAINS

In multidomain grains as in macroscopic ferromagnetic bodies, remanence is induced by irreversible domain-wall movements, that is the domain walls are impelled past energy barriers to new potential minima. Since this process can be thermally activated, it follows that domain walls, and therefore multidomain grains, have blocking temperatures, just as do single-domain grains. At high temperatures the domain walls, being free to move, find the positions of minimum total energy, but they are immobilised by cooling and thus "freeze in" any high temperature (thermoremanent) magnetization. The process of blocking is not as simple as with single domains because domain walls tend to lock into a pattern and the whole pattern must be changed, not a single wall. However, the concept is the same and TRM is a property of all grain sizes.

Now, consider a large spherical grain at its blocking temperature T_B, at which the spontaneous magnetization of its domains is I_{SB}. The domain structure is such that in the absence of an external field the internal flux closure is virtually complete and the grain has no moment. But when a small field H is applied the state of zero moment is not the state of lowest energy. The domain structure is slightly modified to give a magnetization $I \ll I_S$ parallel to H. As was shown by Bhathal et al. (1969) the response of the material of the grain to low fields is isotropic and we should not resolve the field parallel to a preferred domain direction as was previously supposed. Then summing the magnetostatic and field energies:

$$E = \tfrac{1}{2} NI^2 \, V - HIV \tag{7.7}$$

The value of I at T_B is determined by the minimum energy condition, $dE/dI = 0$ (Stacey, 1958), which gives:

$$I = H/N \tag{7.8}$$

This is the condition that the internal field (eq.4.14) is zero. By further cooling the domain walls are constrained to remain in the potential wells which they occupy at the blocking temperature; small reversible displacements in response to low fields are still allowed and contribute to the susceptibility of the grain, as discussed in section 4.2, but irreversible crossing of the barriers can occur only in response to much higher fields. If the spontaneous magnetization of the domains increases to I_S at laboratory temperatures then the resulting thermoremanent moment of the grain is increased by the factor I_S/I_{SB} relative to the value given by eq.7.8. But when placed in a field free space the self-demagnetizing field of the grain $(-NI)$ reduces the magnetization. If the net magnetization is I and the intrinsic susceptibility is χ_i, the reduction in magnetization is $-NI\chi_i$, so that:

$$I = \frac{H}{N} \cdot \frac{I_S}{I_{SB}} - NI\chi_i \tag{7.9}$$

whence:

$$I = \frac{H}{N} \cdot \frac{I_S}{I_{SB}} \cdot \frac{1}{1 + N\chi_i} \tag{7.9}$$

Thus for an assembly of large grains, occupying a volume fraction $f \ll 1$ of a specimen, the TRM per unit volume of the specimen induced in a field H is:

$$M_{TRM} = f \frac{H}{N} \cdot \frac{I_S}{I_{SB}} \cdot \frac{1}{1 + N\chi_i} = H \cdot \frac{I_S}{I_{SB}} \left(\frac{f}{N} - \chi \right) \tag{7.10}$$

In the second form of (7.10), χ is the whole rock susceptibility and χ_i is eliminated by substituting eq.4.17. Eq.7.10 is the equation for multidomain thermoremanence. Everitt (1962a) obtained a different expression which was related to eq.7.10 by Stacey (1963). In effect, Everitt calculated I_S/I_{SB} in terms of the temperature dependences of spontaneous magnetization and coercivity near to the Curie point.

Detailed comparison of eq.7.10 with observations on dispersed magnetite powders in controlled grain sizes is deferred to section 7.5 and 7.6, because, for the lower range of grain sizes of interest, multidomain TRM is masked by the pseudo-single domain effect considered in section 7.4. The observations are compared with the total TRM given by both mechanisms. However, it is useful to indicate at this point that none of the parameters in (7.10) is inaccessible to either observation or theory. Multidomain TRM is only a slight function of grain size and it is convenient to consider 50-μm grains for which $\chi_i = 1.3$ e.m.u., as calculated in section 4.2. The appropriate average value of N for an assembly of randomly oriented grains of elongation about 1.5 : 1, as commonly found in igneous rocks, is $N = 3.9$. Stacey (1958) estimated $I_S/I_{SB} = 3$ from blocking temperatures and spontaneous magnetization curves quoted by Nagata (1953). More recent data show that this estimate was quite good. We now favour the average value $I_S/I_{SB} = 2.5$. So, substituting numerical values, eq.7.10 becomes:

$$M_{TRM} = 0.1_{06} fH \tag{7.11}$$

which is in good agreement with observations on grains larger than about 20 μm.

It is of interest also to compare eq.7.10 with observations by Shive (1969) on TRM in discs of nickel subjected to various amounts of work-hardening. Shive found TRM to increase rapidly with the coercive forces of his specimens. His data showed that the dependence was stronger than a direct proportionality. For the nickel discs $f = 1$, $N = H_c/I_{RS} = 2.2$ (from eq.4.44, since data on H_c and I_{RS} were given), and we assume $\chi_i = 45/H_c$ from eq.4.33. The only parameter of the theory not well controlled for these specimens is I_S/I_{SB}. Reasonable values would be near to unity for the softest specimen (with a low blocking temperature) up to not more than 5 for the hardest disc. We can deduce values of I_S/I_{SB} from H_c and initial gradients of I_{TRM}/H taken from Shive's graphs, using the relation:

$$\frac{I_S}{I_{SB}} = \left(\frac{M_{TRM}}{H}\right) \cdot N\left(1 + N \cdot \frac{45}{H_c}\right) \tag{7.12}$$

The results, given in Table 7.I, coincide with the expected range.

TABLE 7.I

Comparison of Shive's (1969) observations on TRM in discs of work-hardened nickel with the multidomain theory (eq.7.10)

H_c (Oe)	I_{TRM}/H (e.m.u. cm^{-3} Oe^{-1})	I_S/I_{SB} ratio by eq.7.12
32.5	0.45	4.0
25	0.275	3.0
14.5	0.155	2.7
10.5	0.095	2.2
2.3	0.013	1.3

7.4 THE PSEUDO SINGLE DOMAIN (GRAIN SURFACE) EFFECT

The multidomain theory presented in the previous section involves no significantly adjustable parameters and so is readily compared with observations. Comparison soon revealed examples of TRM for which it was seriously in error. There has followed a decade of debate on just what the deficiency is (Verhoogen, 1959; Stacey, 1962a, 1963; Ozima and Ozima, 1965; Dickson et al., 1966). We do not reproduce the arguments here but give our own distillation of what we regard as the theory which has emerged as nearest to correct. In fact it does not reproduce any of the published theories but is in some respects new.

Essentially we are concerned with a compromise between multidomain and single-domain theories — the appearance of some single-domain type properties in multidomain grains. This is called the pseudo-single-domain (p.s.d.) effect. Its importance is a strong function of grain size, being insignificant above about 20 μm but increasingly important at progressively smaller sizes. The reason is simply that multidomain TRM is a grain volume effect but the p.s.d. TRM is a grain surface effect and the grain surface area per unit volume of magnetic mineral is proportional to the reciprocal of grain diameter.

In section 3.5 we showed that the fine structure of closure domains and domain wall edges on the surface of a grain leads to effective single-domain type moments. Since the p.s.d. moments are random effects arising from imperfections in surface domain structures we assume that their magnitudes are uniformly distributed up to a maximum value μ_{max}. The choice of value of μ_{max} is critical to the quantitative application of the theory and at present we can do no better than make a plausible guess. We suppose that the maximum p.s.d. moment corresponds to a rectangular volume of cross-section equal to the critical size for single domains, d_c, and length equal to the domain-wall thickness in bulk material, t, and having an effective magnetization $2I_S/\pi$, as in a domain wall (section 3.5). Thus:

$$\mu_{max} = \frac{2}{\pi} d_c^2 \, t I_S \qquad (7.13)$$

For magnetite (I_S = 480 e.m.u., $d_c \approx 5 \cdot 10^{-6}$ cm, $t \approx 10^{-5}$ cm) this gives $\mu_{max} = 7.6 \cdot 10^{-14}$ e.m.u. Allowing for spaces between them, the average effective surface area occupied by a single p.s.d. moment is $\pi d_c t/2$. Then the number of moments on the surface of a grain is $2 d^2/(d_c t)$ and since the number of magnetic grains of diameter d per unit volume of a specimen containing a volume fraction f of magnetic mineral is $6f/(\pi d^3)$, the total number of p.s.d. moments per unit volume of the specimen is:

$$n = \frac{6f}{\pi d^3} \cdot \frac{2 d^2}{d_c t} = \frac{12f}{\pi d_c \, t d} \qquad (7.14)$$

Now, consider the dn moments with room temperature values in the range μ to $(\mu + d\mu)$:

$$\frac{dn}{n} = \frac{d\mu}{\mu_{max}} \qquad (7.15)$$

These dn moments have a total saturation moment μdn, so that, by eq.7.5, their TRM induced in a field H is:

$$d(M_{TRM}) = \mu dn \int_0^1 x \tanh\left(\frac{\mu_B H}{kT_B} x\right) dx \tag{7.16}$$

where $\mu_B = \mu \cdot \frac{I_{SB}}{I_S}$. This integral is the function $F'(a)$, tabulated in Appendix 1. For small fields, 7.16 integrates to:

$$d(M_{TRM}) = \frac{\mu^2 H}{3kT_B} \cdot \frac{I_{SB}}{I_S} \cdot \frac{n}{\mu_{max}} d\mu \tag{7.17}$$

The total low field TRM of all p.s.d. moments in the specimen is obtained by integrating with respect to μ from 0 to μ_{max}:

$$M_{TRM} = n \frac{\mu_{max}^2 H}{9kT_B} \cdot \frac{I_{SB}}{I_S} = \frac{16}{3\pi^3} \frac{(d_c^3 t I_S^2)}{kT_B} \left(\frac{I_{SB}}{I_S}\right) \frac{fH}{d} \tag{7.18}$$

Substituting numerical values for magnetite grains:

$(d_c = 5 \cdot 10^{-6}$ cm, $t = 10^{-5}$ cm, $I_S = 480$ e.m.u., $I_{SB}/I_S = 0.4$, $T_B = 750°K$):

$$\frac{M_{TRM}}{fH} = 1.9 \cdot 10^{-4}/d \quad (d \text{ in cm}) \tag{7.19}$$

As shown in Fig. 7.1 this agrees reasonably well with the low-field TRM of specimens with grain diameters below about 20 μm. Dunlop (1973b) has recently extended TRM measurements on grains of controlled sizes to the range 0.037 μm to 0.22 μm and found its strength to depend slightly less strongly upon d than eq.7.19 over the range between his data and those of Parry (1965). Since a diminution in μ_{max} is expected at the smallest sizes, this observation is a valuable confirmation of the essential validity of the theory. However, the agreement should not conceal the arbitrariness in some of the numerical values used to obtain eq.7.19. Particularly since d_c enters eq. 7.18 as a cube it is evident that any error in estimating its value is greatly amplified in the deduced value of I_{TRM}. Apparently the best justification for the value assumed is that it gives a thermoremanence in accord with observations. However, further observational checks are highly desirable and the most obvious way of checking the upper limit μ_{max} in the range of p.s.d. moments is to look for saturation effects, i.e., a non-linearity of TRM with field, even in moderately low fields. The theory presented here is very vulnerable to this particular test because a specific and simple size distribution of p.s.d. moments is assumed and a precise shape of the I_{TRM} vs H curve is therefore implied, independently of grain size. The curve must be obtained numerically by integration of eq.7.16 with respect to x for each of a series of values of $(\mu_B H/kT_B)$ and then with respect to μ for each of a series of values of H. The result is shown as Fig. 7.2, which uses the tabulation in Appendix 2 of the function:

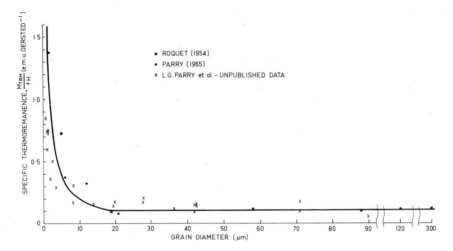

Fig.7.1. Low field thermoremanence per oersted per unit volume fraction of magnetic mineral as a function of grain size.
(Data by Roquet, 1954, and Parry, 1965, compared with the theoretical relationships expected from eq.7.11 – $d > 18\ \mu m$ – and 7.19 – $d < 18\ \mu m$. This curve is similar to one by Dickson et al., 1966.)

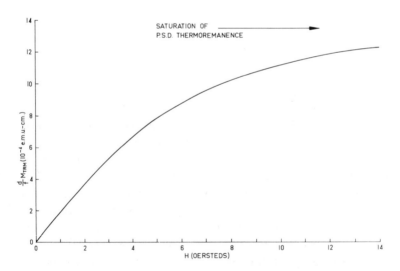

Fig.7.2. Pseudo-single domain contribution to thermoremanence as a function of inducing field H for a sample containing a volume fraction $f \ll 1$ of grains of diameter $d < 20\ \mu m$, according to eq.7.21.

$$F(a) = \int_0^1 \int_0^1 xy \tanh(axy)\, dx\, dy \qquad (7.20)$$

Eq.7.16 integrates to give:

$$M_{TRM} = n\mu_{max} F\left(\frac{\mu_{max}H}{kT_B} \cdot \frac{I_{SB}}{I_s}\right) \qquad (7.21)$$

and with the numerical values for magnetite assumed here:

$$\frac{\mu_{max}}{kT_B} \cdot \frac{I_{SB}}{I_S} = 0.30/\text{Oe} \qquad (7.22)$$

and:

$$n\mu_{max} = 5.8 \cdot 10^{-3} \, (f/d) \qquad (7.23)$$

Eq.7.18 and 7.19 give the initial gradient of $I_{TRM}(H)$ as represented by 7.21. An experimental value of 0.28 was obtained by D.E.W. Gillingham for the parameter in (7.22), by fitting eq.7.21 to very accurate TRM data for a basalt (see Fig.7.3). This agreement is well within the uncertainties of the present estimates.

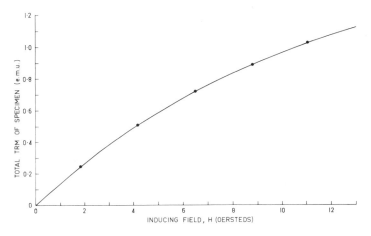

Fig.7.3. TRM versus inducing field strength for a specimen of basalt from Mt. Tamborine, Queensland. (Data by Gillingham, 1971, compared with a theoretical curve for a specimen in which the bulk of the magnetite — 0.22 cm³ — is assumed to give multidomain TRM linear in field by eq.7.11 and a p.s.d. contribution by eq.7.21 is added but with the fitted parameter $(\mu_{max}/kT_B)(I_{S_B}/I_S) = 0.28$.)

7.5 TOTAL THERMOREMANENCE

In section 3.5 we presented a model of a magnetite grain as a volume of material which is magnetizable by the normal laws of bulk ferromagnetics and a surface whose properties are quite different, being similar to an assembly of single domains. The volume of the grain acquires thermoremanence according to eq.7.10 and the surface according to eq.7.18 or 7.21. The remaining problem is to decide how these two effects superimpose.

Consider first a small grain in which the surface effect is dominant. We have on the grain surface a skin of material which, at the blocking temperature, is more highly magnetizable than the bulk of the grain. As pointed out in section *3.5*, provided the skin is uniformly magnetized it exerts no *net* field on the interior of the grain, but being more highly magnetizable it acts as a magnetic screen and prevents an external field from magnetizing the bulk of the grain. Thus for small grains eq.7.18 applies and 7.10 does not. The upper limit of the size range for which this is so is the size for which the two equations give equal TRM. Numerical values for magnetite (eq.7.11 and 7.19) give this size as 18 μm. For larger sizes the volume of the grain would acquire a higher magnetic moment than the skin which is screening it. The effect of the screening therefore diminishes with increasing grain size and the sum of p.s.d. and multidomain TRM is simply the TRM which would be given by eq.7.10 alone. The resulting plot of TRM vs grain size (Fig. 7.1) shows very strikingly a discontinuity in properties at about 18 μm.

Rocks normally contain a wide range of grain sizes, so that both multidomain and p.s.d. effects appear. It is only from data on controlled grain sizes that the distinction becomes apparent. The measurements reported by Parry (1965) are of particular value in this connection. Measurements of the field dependence of TRM in sub-20μm grains to check eq.7.21 and probably at least adjust the parameters are needed. However, working with mixed grain sizes one still has a check on the form of the curve in Fig.7.2, by assuming a mixture of true multidomain and p.s.d. TRM. The result of such a check is shown in Fig.7.3. This is a sample of data obtained by D.E.W. Gillingham using a spinner magnetometer with a ratio-transformer potentiometric measurement of output voltage to give very high precision on the magnitude of measured moments. He also had a very fine control on the currents in his magnetizing solenoid. He has found no rock sample in which TRM is perfectly linear in field over the range shown. There is always some curvature, indicating a p.s.d. contribution to TRM.

Lowrie and Fuller (1971) reported that the resistance to alternating field demagnetization of TRM in multidomain magnetite increased with the strength of the field in which TRM was induced. They argued that this allowed multidomain TRM to be distinguished from single-domain TRM, whose resistance to demagnetization decreased with increasing strength of the inducing field. However, their result is not a general one. Experiments by D. E. W. Gillingham at the University of Queensland on carefully sized and dispersed multidomain magnetite (in the size range 100–200 μm) have shown that high field TRM may be either more or less resistant to demagnetization and no example as striking as those reported by Lowrie and Fuller (1971) has been found. Some superimposed demagnetization curves for high- and low-field TRM have been found to cross over, the high field TRM being less resistant to demagnetization in low alternating fields and more resistant in high alternating fields. However, Gillingham's observations are subject to the same general explanation as Lowrie and Fuller gave for theirs. More intense TRM requires more (or bigger) domain wall displacements, so that there are more potential barriers to be crossed by domain walls in demagnetizing it. In a multidomain grain the domain walls which move most easily move first and for very low-field TRM they may suffice to demagnetize the grain, but for complete demagnetization of high-field TRM it may be necessary to move domain walls (or cause domain rotations) past higher barriers.

Probably the most satisfactory method of distinguishing single-domain or p.s.d. thermoremanences from multidomain thermoremanence is the resistance to demagnetization by cooling. As Merrill (1970) and L. G. Parry (personal communication, 1971) have shown multidomain TRM in magnetite grains is destroyed by cooling through the isotropic point (118°K) but p.s.d. TRM is recovered on re-warming.

7.6 THE KOENIGSBERGER RATIOS

The strength of thermoremanence in a specimen, relative to the quantity of magnetic mineral, is expressed in terms of a parameter Q_t, defined by Koenigsberger (1938) as the ratio of thermoremanence to the magnetization induced at laboratory temperature by the same field:

$$Q_t = \frac{M_{TRM}}{\chi H} \tag{7.24}$$

A similar ratio which has less fundamental significance but more practical value in paleomagnetism is:

$$Q_n = \frac{M_{NRM}}{\chi H} \tag{7.25}$$

H being the present earth's field strength and M_{NRM} is the natural remanence of a specimen. If the natural remanence is thermoremanent in origin, and if it was induced in a field of strength similar to the present field, Q_n and Q_t are comparable.

For multidomain grains, with M_{TRM} given by eq.7.10 and χ by 4.17 the expression for Q_t is very simple:

$$Q_t = \frac{1}{N\chi_i} \cdot \frac{I_S}{I_{SB}} \approx 0.5 \tag{7.26}$$

(with numerical values for magnetite grains of 50 μm diameter: $N = 3.9$, $\chi_i = 1.3$ e.m.u., $I_S/I_{SB} = 2.5$). The maximum plausible range for multidomain grains is $0.2 \leqslant Q_t \leqslant 1.0$. However, grains with pseudo-single domain properties have stronger thermoremanence and hence higher Q_t values, as is shown clearly by the data of Table 7.II. The effect is so striking that a measurement of Q_n suffices to show whether the remanence of a rock has a substantial p.s.d. component; thermoremanence experiments are not necessary.

Practical interest in Q_n (or Q_t) is that a high value ($\gg 0.5$) indicates the presence of fine grains which dominate the remanence of a specimen and being fine they have

TABLE 7.II

Values of Koenigsberger ratio Q_t (eq.7.21) for laboratory-prepared specimens with a wide range of sizes of dispersed magnetite grains. (Values of I_{TRM} and χ are from Parry, 1965, and the Q_t values were quoted by Stacey, 1967a).

Grain diameter (μm)	I_{TRM} in 0.4 Oe (e.m.u. per cm^3 of magnetite)	χ (e.m.u. per cm^3 of magnetite)	Q_t
120	0.044	0.24	0.46
88	0.041	0.21	0.49
58	0.046	0.22	0.52
21	0.032	0.21	0.38
19	0.041	0.19	0.54
6	0.15	0.19	2.0
1.5	0.55	0.19	7.2

high coercivities and, almost necessarily, high stabilities. Thus from the point of view of paleomagneticists the merit of high Q_n is that it indicates magnetic stability.

7.7. PARTIAL THERMOREMANENCES AND THE LAW OF ADDITIVITY

A specimen will acquire a partial thermoremanence if it is cooled in a field through a limited temperature range, being cooled in zero field through the remaining ranges. Those grains whose blocking temperatures are within the range of field cooling acquire thermoremanence and the others do not, the resulting magnetization of the specimen as a whole being known as a partial thermoremanent magnetization (PTRM). E. Thellier first noticed many years ago that the PTRM's acquired in different temperature intervals are independent (see, for example, Thellier and Thellier, 1941). This means that the total TRM acquired by cooling a specimen in a field from the highest Curie point of its magnetic minerals to laboratory temperature is equal to the sum of the PTRM's acquired separately in several temperature intervals which together make up the whole temperature range. This law of additivity is well established by an extensive literature (e.g., Nagata, 1961). It follows naturally from the theory of TRM in independent single-domain grains, each of which may be considered to acquire TRM at its blocking temperature independently of the other grains. The reason for its quite general applicability is less obvious; some recent work throws new light on this and on the blocking temperature mechanism in multidomain grains.

PTRM spectra (Fig.7.4) show clearly the important range(s) of blocking temperature in any specimen. The spread is normally several tens of degrees even in simple, single-constituent materials and may be much wider. The presence of two or more apparently separate constituents may be indicated, as in Fig.7.4(b). A PTRM acquired in a temperature interval T_1 to T_2 ($T_1 > T_2$) is hardly affected by re-heating to T_2 and virtually

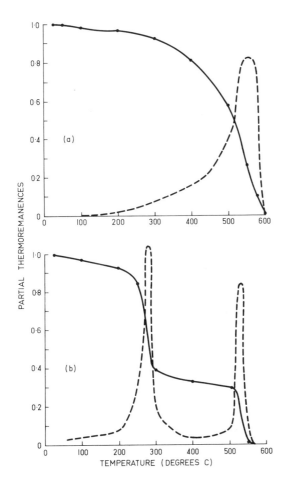

Fig.7.4. Partial thermoremanence spectra of two basalt specimens (a) having a single constituent, (b) having two constituents with different blocking temperature ranges. The solid curves join experimental points showing values of PTRM obtained by cooling specimens from the indicated temperatures to laboratory temperature. Broken curves are approximate differentials of the solid curves and show the PTRM acquired per unit temperature interval.

destroyed by further heating to T_1 and subsequent cooling in zero field, so that integrated PTRM spectra (dashed curves in Fig.7.4) are also thermal demagnetization curves.

Stacey (1963) showed that, with certain assumptions, the law of additivity applied to multidomain grains having different blocking temperatures in different parts. He considered the interaction fields between two parts of a single grain with different blocking temperatures and allowed for the mutual demagnetization in calculating both PTRM's and total TRM. However, a more detailed calculation of this model, with an allowance for intrinsic susceptibility (which varies with temperature), fails to give perfect additivity. Further, this model predicts that additivity will break down when the inducing field is high enough to give a total TRM non-linear in field. D. E. W. Gillingham has tested this prediction and found that for a number of rocks TRM becomes quite strongly non-linear in fields up to 50 Oe, but that PTRM's in these fields are nevertheless still additive, within the uncertainties of measurement. We, therefore, conclude that the model of a multi-domain grain with different blocking temperatures in different parts is invalid. For the grain sizes of interest in rock magnetism it appears that the whole domain structure of a grain has the same blocking temperature i.e., all parts are interdependent to some extent. Additivity is then independent of field strength or linearity of TRM, as observed, because the components with different blocking temperatures are different grains and are therefore not interacting with one another; no difficulty then arises with the susceptibility correction of TRM:

7.8 THERMOREMANENCE IN ANISOTROPIC ROCK

The theory of thermoremanence in an anisotropic rock containing only multi-domain grains follows directly from eq. 7.10, if the magnetic grains are represented by a single equivalent ellipsoid. This is justified by the conclusion reached in sections 4.3 and 7.3 that in low fields the material in large grains is intrinsically isotropic in behaviour. Applying a field H at an angle θ to the a axis of an ellipsoid, as in Fig.7.5, we obtain the TRM in a and b directions and hence the angle ϕ of TRM:

$$M_a = f \frac{H \cos \theta}{N_a} \cdot \frac{I_S}{I_{SB}} \cdot \frac{1}{1 + N_a \chi_i} \quad (7.27)$$

$$M_b = f \frac{H \sin \theta}{N_b} \cdot \frac{I_S}{I_{SB}} \cdot \frac{1}{1 + N_B \chi_i} \quad (7.28)$$

so that:

$$\tan \phi = \frac{M_b}{M_a} = \frac{N_a (1 + N_a \chi_i)}{N_b (1 + N_b \chi_i)} \tan \theta = X \tan \theta . \quad (7.29)$$

From the point of view of paleomagnetism we are interested in the maximum possible deflection of thermoremanence by a slight anisotropy. Thus:

$$\tan (\theta - \phi) = \frac{\tan \theta - \tan \phi}{1 + \tan \theta \tan \phi} = \frac{(1 - X) \tan \theta}{1 + X \tan^2 \theta} \quad (7.30)$$

and the condition for $(\theta - \phi)$ to be a maximum is $d(\theta - \phi)/d\theta = 0$ which gives:

$$\tan \theta = X^{-1/2} \quad (7.31)$$

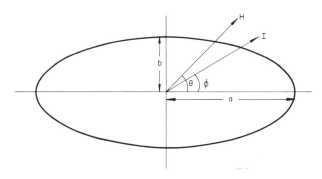

Fig.7.5. Equivalent ellipsoid of an anisotropic rock with multidomain grains, magnetized at an angle ϕ to its major axis by a field at an angle $\theta > \phi$.

and therefore:

$$\tan \phi = X^{1/2} \tag{7.32}$$

If we specify that a 3°-deflection is the maximum normally acceptable in paleomagnetism, this being the approximate error of measurement in careful work, then θ, ϕ each depart from 45° by 1.5° and $(X^{-1/2})_{max} = 1.05$ or:

$$\left[\frac{N_b(1+N_b\chi_i)}{N_a(1+N_a\chi_i)}\right] = 1.11 \tag{7.33}$$

This expression must be translated into measurable expressions for anisotropy. First, noting eq.4.17, we can reduce eq.7.33 to:

$$\left[\frac{N_b}{N_a} \cdot \frac{\chi_a}{\chi_b}\right]_{max} = \left[\left(1 + \frac{\Delta N}{N}\right) \cdot \left(1 + \frac{\Delta \chi}{\chi}\right)\right]_{max} = 1.11 \tag{7.34}$$

In high field anisotropy experiments we measure $(N_b - N_a)$ and in low field measurements $(\chi_a - \chi_b)$. These two measures of anisotropy are related by eq.4.41, which, for small anisotropies, is:

$$\frac{\Delta N}{N} = \frac{5}{N}\frac{\Delta \chi}{\chi} = 1.28 \frac{\Delta \chi}{\chi} \tag{7.35}$$

from which:

$$\left(\frac{\Delta \chi}{\chi}\right)_{max} = 0.048 \tag{7.36}$$

$$(N_b - N_a)_{max} = 0.24 \tag{7.37}$$

and, by eq.3.8:

$$\left(1 - \frac{b}{a}\right)_{max} = \frac{5}{8\pi}(N_b - N_a)_{max} = 0.048 \qquad (7.38)$$

Thus without the application of a correction the tolerable anisotropy, in terms of either susceptibility or shape elongation of the grains, is 5 %, as reported by Uyeda et al. (1963). However, a simple correction is applied automatically in the process of partial demagnetization by an alternating field to remove unwanted secondary components of natural magnetization. Demagnetization is more effective in the easy direction of magnetization because the applied alternating field is opposed by a smaller self-demagnetizing field. This process is considered further in Chapter 10.

The effect of anisotropy on **pseudo-single domain thermoremanence** is less clear. At first sight it would appear that, being a surface effect, the p.s.d. TRM would be unaffected by grain elongations, but that would assume that the p.s.d. moments are individually unaffected by the orientation of the grain surface on which they are situated. We cannot assume this to be so. However, as we have neither theory nor observations of anisotropy of p.s.d. TRM, we merely note this as a problem requiring attention. But if the criterion of acceptable anisotropy by the foregoing multidomain theory is adopted then it appears to suffice also for p.s.d. TRM.

Chapter 8

DEPOSITIONAL REMANENT MAGNETIZATION·(DRM)

8.1 THE OCCURRENCE OF DRM

When magnetized grains are aligned by the earth's field during sedimentation, the resulting sediment acquires a remanence with a stability characteristic of the original process of magnetization of the grains. Thus grains of magnetite which originated in an igneous rock, in which they acquired thermoremanence, may impart to a sediment in which they are deposited a DRM which has a stability similar to that of the TRM, although it is less intense. The significance of DRM was established by Johnson et al. (1948), who examined the paleo-secular variation of the geomagnetic field apparent in the directions of remanence of New England varved (layered) clays. They re-deposited clay in controlled laboratory fields and obtained the field dependence of DRM in Fig.8.1, showing that the natural remanence could be interpreted as DRM acquired during deposition in a field comparable to that of the present geomagnetic field. The magnetization was attributed to magnetite in extremely small grain sizes, estimated to be 0.3 μm or smaller.

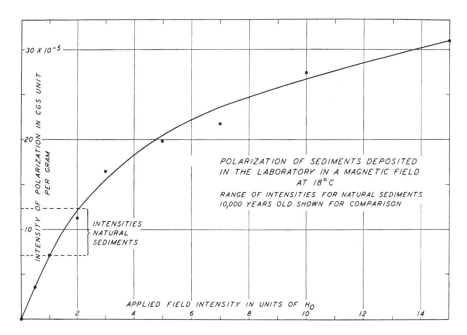

Fig.8.1. Intensity of detrital remanence as a function of field strength during laboratory redeposition of a varved clay sample from New England.
(Reproduced by permission, from Johnson et al., 1948.)

Subsequent work in Britain and Sweden has generally confirmed the conclusion of Johnson et al. (1948) that DRM is normally dominated by magnetite grains, but Collinson (1965a) found that the magnetization of a Triassic sediment in U.S.A. was due to detrital, black hematite grains. However, in many cases the magnetizations of sediments are not detrital but chemical in origin, having been produced some time after deposition, during chemical alteration in the process of consolidation. In this process magnetite is converted to hematite via a hydrated oxide. Thus DRM is virtually restricted to recent sediments which have not suffered any chemical, consolidation process.

8.2 THE DETRITAL MAGNETIZATION PROCESS

As in Fig.8.1, laboratory experiments on DRM have shown that its strength increases with field strength, although the detailed field dependence varies between samples. Thus Khramov (1968) found DRM linear in field up to at least 5 Oe, but the specimens of Johnson et al. (1948) showed a strong tendency to saturation alignment in a similar field. The problem of the field dependence has been the subject of diverging opinions in the literature. The approach taken here follows Stacey (1972).

The magnetic torque exerted on a grain of magnetic moment m by a field H at an angle θ to m is:

$$L = -mH \sin \theta \qquad (8.1)$$

Since moments of inertia vary as d^5, rotational accelerations of very small grains may be neglected; the equilibrium rate of grain rotation results from the balance of magnetic torque and the opposing torque of viscous drag:

$$-mH \sin \theta - \pi d^3 \eta \frac{d\theta}{dt} = 0 \qquad (8.2)$$

where η is the viscosity of water, whence:

$$\frac{d\theta}{dt} = -\frac{mH}{\pi d^3 \eta} \sin \theta = -\frac{IH}{6\eta} \sin \theta \qquad (8.3)$$

since $m = \pi d^3 I/6$ for a spherical grain of magnetization I per unit volume. For small angles θ, eq.8.3, gives an exponential decay of θ with a time constant:

$$\tau = \frac{\pi d^3 \eta}{mH} = \frac{6\eta}{IH} \qquad (8.4)$$

which is therefore characteristic of the rate of alignment of grains in a field. With plausible numerical values, $\eta = 10^{-2}$ poise, $H = 0.5$ Oe, $I = 0.1-10$ e.m.u./cm^3, we find $\tau =$ = 1.2–0.012 sec, so that magnetite grains would very quickly assume perfect alignment.

This is emphasized by considering the depth h of water through which they fall in time τ', which is by Stoke's law:

$$h = \frac{d^2(\rho - \rho_0)g\tau'}{18\eta} \tag{8.5}$$

$(\rho - \rho_0) \approx 4$ g/cm^3 being the density difference between the grains and the water and $g = 981$ cm/sec^2 is the gravitational acceleration. For grains of diameter $10\,\mu$m (10^{-3} cm) and $\tau' = 1.2$ sec we obtain $h = 0.026$ cm, i.e., even for large, relatively weakly magnetized grains, alignment becomes complete in the earth's field during settling in extremely shallow water. Equating τ by eq.8.4 with τ' and substituting in eq.8.5 we obtain the general relationship between h and d:

$$h = \frac{d^2(\rho - \rho_0)g}{3IH} \tag{8.6}$$

Since the field dependence of DRM shows the magnetic grain alignment to be incomplete in small fields, and the depth of water is clearly irrelevant, another misaligning influence must be sought. Collinson (1965b) suggested the Brownian motion of small magnetite grains. Under the influence of thermal agitation the equilibrium alignment of grains of magnetic moment m yields a specimen magnetic moment M which is a fraction of that due to complete alignment, M_0, given by the Langevin formula for a classical paramagnetic at absolute temperature T:

$$M = M_0 \left[\coth\left(\frac{mH}{kT}\right) - \left(\frac{kT}{mH}\right) \right] \tag{8.7}$$

where k is Boltzmann's constant (see Appendix 3). For alignment to be much less than saturation we require $mH/kT < 1$ and, at $T = 290°$K with $H = 0.5$ Oe and I in the range $0.1-10$ e.m.u./cm^3 as considered above, this means an upper limit for grain diameter in the range $1-0.25\,\mu$m. For smaller grains (or weaker grain remanence, or weaker magnetic field) eq.8.7 reduces to:

$$\frac{M}{M_0} = \frac{mH}{3kT} \tag{8.8}$$

in which range the magnetic alignment is linear in the inducing field. For $mH/kT > 3$ the approach to saturation alignment becomes obvious.

There can be little doubt that Brownian motion provides a qualitatively adequate explanation of the partial alignment of grain moments in DRM, but quantitatively eq.8.7 is deficient in that it gives much too sharp a saturation to explain directly observations such as those of Johnson et al. 1948 (Fig. 8.1). The effectiveness of Brownian motion is necessarily limited to a small range of grain sizes such that mH/kT is of order unity. King and Rees (1966) doubted its applicability to the theory of DRM for this reason, asserting that the majority of grains show either very little alignment or else virtually saturated alignment. However, by considering a distribution of grain sizes, the range of sizes over which alignment is effective increases with the field strength and the resulting DRM vs field curve becomes a much better approximation to the experimental data. Stacey (1972) considered a uniform distribution of grain moments up to a maximum value m_{max}, that is each range dm up to m_{max} contributes to the moment at saturation alignment an amount proportional to dm. With this assumption, eq.8.7 integrates over the range of m to give:

$$\frac{M}{M_0} = \frac{1}{x} \ln\left(\frac{\sinh x}{x}\right) \tag{8.9}$$

where $x = m_{max} H/kT$. For small x, i.e., for an assembly of grains for which $m_{max} < kT/H$, (8.9) reduces to:

$$\frac{M}{M_0} = \frac{m_{max} H}{6 kT} \tag{8.10}$$

Eq.8.7 and 8.9 are plotted in Fig. 8.2 in which the initial gradients have been matched by dividing the Langevin expression (8.7) by 2 so that the approach to saturation

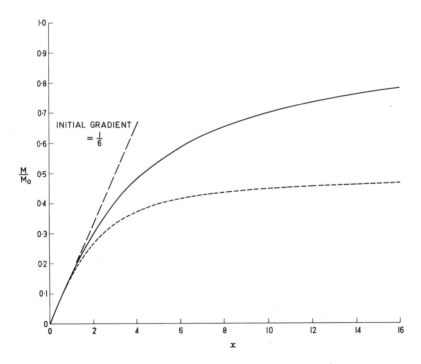

Fig.8.2. Plots of M/M_0 versus $x = m_{max} H/kT$ according to eq.8.9 (solid line) and $\frac{1}{2} M/M_0$ by eq.8.7 (broken line). The scale has been matched to give an approximate fit of the upper curve, representing an assembly of grains with a range of magnetic moments up to m_{max}, to the curve of Fig.8.1. The lower curve, representing grains of uniform size, saturates much too sharply to give a satisfactory representation of the experimental data.

is easily compared. Also eq.8.9 is tabulated in Appendix 4. The similarity of the solid curve, representing eq.8.9, to the data in Fig.8.1 is sufficiently good to justify the calculation of grain moments for the samples of Johnson et al. (1948) by matching the two curves. Stacey (1972) obtained $m_{max} \approx 7.4 \cdot 10^{-14}$ e.m.u. He further showed that since values of saturation remanence M_{RS} and coercive force H_c were also given for this material and these two quantities are related by eq.4.44 or better still by the equivalent experimentally deduced equation of Parry (1965):

$$M_{RS} = fI_{RS} = f \times 0.30 H_c \tag{8.11}$$

the value of the volume fraction of magnetic mineral $f = 1.0 \cdot 10^{-4}$ could be deduced directly and that this led to an estimate of the magnetization per unit volume for the natural magnetizations of the grains. The value obtained, 8 e.m.u./cm^3 has an important bearing on the theory of magnetic properties of fine grains because it is too large by a factor ~ 100 to be compatible with multidomain remanence but too small by a factor ~ 50 for single-domain magnetization. It is a convincing demonstration of the existence of pseudo-single domain moments in very fine multidomains (sections *3.5* and *7.4*).

In the case of the varves examined by Johnson et al. (1948) the upper limit of magnetite grain size appears to have been about 0.25 μm and the corresponding value of m_{max} was such that $m_{max} H/kT \approx 1$ in the earth's field, making it ideal material for a test of the theory. However, many of the sediments which have been used in laboratory deposition experiments have much larger grains, making the applicability of Brownian motion doubtful. Some alternative processes which reduce the intensity of DRM are considered in section *8.4*.

8.3 DEMAGNETIZATION AND STABILITY OF DETRITAL REMANENCE

The stability of DRM and its resistance to demagnetization are merely consequences of the original mechanism of magnetization of the grains responsible for DRM. However, the curve of alternating field demagnetization of DRM is not, in general, the same as that of TRM in the same sample because the different grain sizes contribute in different proportions to the two kinds of remanence. The only experiments directly concerned with this problem appear to be by Rusakov (1967), whose results refer mainly to hematite. In this respect they are probably of greater interest in connection with the properties of hematite grains than with DRM itself, the somewhat surprising conclusion being that DRM is consistently more resistant to alternating field demagnetization than TRM, and under thermal demagnetization it is more resistant up to 300°C. Our explanation for this differs from that given by Rusakov. The magnetic properties of hematite grains are highly variable between grains, being controlled largely by crystal defects. This results in a correlation between magnetic hardness and the strength of magnetization of individual grains. In DRM the more strongly magnetized (and therefore harder) grains contribute proportionally more to the total magnetization because of their greater degree of alignment, and DRM is therefore harder. Rusakov's (1966) data on magnetite is concerned with the close similarity between alternating field demagnetization curves for DRM and ARM. However, he used grains in the size range 80–100 μm, whose moments are such that the misaligning effects of Brownian motion would be insignificant and alignment close to saturation is expected. This conclusion is confirmed by Rusakov's report that ARM induced in magnetic samples by a steady field of 0.4–0.5 Oe was equal to the detrital remanence induced in the earth's field (which is similar in magnitude). Thus the grains contribute similarly to both kinds of remanence, which cannot be distinguished by their demagnetization curves. This work tells us very little about DRM in varves with very fine grained magnetite.

8.4 INCLINATION ERRORS, WATER FLOW AND OTHER EXTRANEOUS EFFECTS

One of the problems which have arisen in the use of detrital remanence in paleomagnetism is the occurrence of inclination errors: DRM is nearer to horizontal than the field in which it is produced. Of the explanations which have been advanced, two appear to be particularly important as they affect not only the inclination of DRM but also its intensity and so have a direct bearing on the discussion in section 8.2.

King (1955) suggested that as grains tend to settle with long axes horizontal and the magnetic anisotropy of an individual grain favours magnetization in the direction of a long axis, so a net deflection of DRM towards the horizontal is produced. His simple numerical model of the process envisaged an assembly of grains of two kinds, a fraction f being spheres which settle with no inclination error and $(1 - f)$ being discs which settle horizontally, so that they can contribute only to the horizontal component of magnetization. Then by supposing that the magnetization resulting from alignment of the spheres is proportional to field strength, and alignment of the discs to the horizontal component of the field, he obtained the relationship between the inclinations of DRM, θ_I and field, θ_H:

$$\tan \theta_I = f \tan \theta_H \tag{8.12}$$

The assumption of proportionality to field is important to the argument because if saturation alignment is assumed instead then the dip angle becomes independent of the field, being determined only by the proportions of spheres and discs. Thus King (1955) implicitly supposed a randomizing mechanism, such as Brownian motion with $mH/kT < 1$, to operate, although at least some of his grains appear to have been too large.

If King's (1955) argument is basically valid, and the evidence of magnetic anisotropy of sediments clearly indicates horizontal alignment of long axes, then there is another consequence. By recognizing that hydrodynamic torques acting on a precipitating grain exercise a control on its orientation, and that the average dimension ratio of grains is only of order 1.5 : 1 so that the preference of magnetic moments for the long axes is not very strong, we introduce a randomizing influence on the magnetic moments, independently of any Brownian motion.

We must accept that an inclination error in sediments is associated with horizontal layering and can be regarded as a consequence of the magnetic anisotropy, but to match his observations on laboratory re-deposition of Swedish varves, King (1955) required the fraction of spheres in his mixture to be as low as 0.4, which appears implausible. This led to the consideration of another mechanism, which was discussed more rigorously by Griffiths et al. (1960), who considered spherical magnetic grains which were oriented in the field direction as they approached the bottom of the water but rolled into the nearest hollow in the surface formed by the previously deposited grains. If it is supposed that the grains are rotated about randomly oriented horizontal axes by a common angle ϕ then the resulting remanence is found to be deflected as by eq.8.12, where:

$$f = \frac{2 \cos \phi}{1 + \cos \phi} \qquad (8.13)$$

For $f = 0.4$ we require $\phi = 75°$. But analysis in terms of a single angle of rotation is too simple; grains rotated by a large angle will, because of the wide dispersion of their moments, contribute little to the resulting remanence, which will be dominated by the grains which have rotated through small angles. Thus the required average rotation is substantially larger than 75° which, again, appears implausible. It therefore appears necessary to suppose that both of the mechanisms considered operate together. We can also note that the grain rolling process is another mechanism for magnetic grain misalignment and consequent reduction in DRM in the final sediment.

Although the inclination error is a well documented problem in laboratory redepositions and is believed to apply to glacial varves, it is by no means general. In many cases, notably marine sediments, it does not appear to occur. The possibility has been noted (Harrison, 1966) that burrowing organisms so re-work the top few centimetres of sediment that alignment during the deposition process itself becomes irrelevant. In any case it is known that field alignment can occur some time after deposition if a sediment remains wet enough (Irving and Major, 1964; Khramov, 1968).

Two further effects are related to the problem of inclination errors. A bedding error, due to deposition on an inclined surface follows immediately from the grain rolling idea, if grains roll downhill through larger angles, on average, than uphill (Hamilton and King, 1964). The other problem is that of water flow. Rees (1961) found that his measurements were explicable in terms of a simple equilibrium rotation of the individual grains due to the combination of magnetic torque and viscous torque due to velocity gradient of the water. It must be supposed that turbulent water introduces a randomness, but this appears unimportant as the thin boundary layer of water at the bottom suffices to allow small grains to come to equilibrium orientation.

Chapter 9

CHEMICAL REMANENT MAGNETIZATION (CRM)

9.1 THE PALEOMAGNETIC SIGNIFICANCE OF CRM

When a magnetic material is produced by a chemical process or phase change at a temperature below its Curie point, the remanence which it thus acquires in the ambient field is referred to as Chemical Remanent Magnetization. This is a common natural occurrence, so that the nature of CRM is of considerable importance to paleomagnetism. The principle has been recognized for over 100 years, but little recent quantitative work, either experimental or theoretical, has been done on it and it is the least understood of the processes by which rocks and minerals become magnetized.

The first relevant laboratory experiment appears to have been by W. Beetz (1860) (see Ewing, 1914) who electrolytically deposited iron on to a silver wire in a magnetic field and found it to be so strongly magnetized that the subsequent imposition of a much stronger field had little further effect. Beetz thus discovered the two essential features of CRM: its intensity is much greater than that of isothermal magnetization in comparable small fields, and it is more stable. The suggestion that CRM could be important in rocks was made by Koenigsberger (1938, p.319), who noted the very strong and stable remanences of certain sediments, which could have been either detrital remanence (Chapter 8) or "crystallization remanence" resulting from the formation of their magnetic minerals after deposition. It is still by no means always clear whether the remanences of sediments are detrital or chemical in origin.

The obvious physical situation favouring CRM is found in red sandstones in which the hematite present appears to have been formed by oxidation of the originally deposited magnetite via hydrated oxides. In sandstones, in which the remanence is associated specifically with the hematite coatings on quartz grains, it is clearly chemical in origin, but Collinson (1965a) found red sandstones in which the remanence was carried by black hematite grains and was virtually unaffected by dissolving away the red coatings. Since igneous rocks contain titanomagnetites but not hematite, it is apparent that the hematite which is common in sediments is not normally detrital but is produced in situ chemically, so that CRM is certainly common and probably dominates the magnetizations of all but the most recent sediments. From the paleomagnetic point of view this is perfectly satisfactory if the oxidation follows the deposition of a sediment sufficiently rapidly that the CRM still effectively dates from the deposition itself. Bagina (1966) described a rock in which the superimposed magnetizations of magnetite grains and of secondary hematite were distinguished by progressive thermal demagnetization. The indicated declinations coincided, but it appeared that the magnetite gave a shallower inclination, indicating that its contribution of the magnetization was detrital in origin and that the hematite faithfully recorded the field direction at the time of deposition.

Recognition that CRM may be very important also in igneous rocks is more recent. Conventionally we think of them as acquiring thermoremanence by the mechanisms discussed in Chapter 7; in rapid laboratory cooling of rocks this must normally be so and probably also in igneous rocks which originally cooled very fast and have not been subjected even to moderate re-heating. However, historical lava flows have been known to have temperatures of the order of 100°C years after their eruption. More importantly, high-temperature ($\sim$ 900°C–1000°C) and low-temperature ($<$ 600°C) oxidation of basalts were shown by Buddington and Lindsley (1964) to be the primary origins of the exsolved ilmenite lamellae commonly present in the titanium-poor magnetites in basalts. Ilmenite is paramagnetic at room temperature and so is unable to acquire TRM or CRM, but the accompanying titanium-poor magnetite will do so. If it is produced by high-temperature oxidation above its Curie point of 580°C, then it will acquire TRM with further cooling. Carmichael and Nicholls (1967) have pointed out that high-temperature oxidation is often accompanied by pseudo-brookite (Fe_2TiO_5) which is unstable below 600°C so that its presence supports a thermoremanent origin for the observed NRM. If, on the other hand, low-temperature oxidation below the Curie point of the titanium-poor magnetite took place a CRM will result. Using magnetic measurements down to the Néel point of the exsolved hemo-ilmenites, Grommé et al. (1969) showed unmistakable presence of low-temperature oxidation in quenched basalt samples obtained from two lava lakes of Kilauea volcano, Hawaii. High-temperature quenching (from above 800°C) showed Curie points of single-phase titanium-rich magnetite whereas low-temperature quenching (from 700°C down to 40°C) resulted in an additional Curie point corresponding to exsolved titanium-poor magnetite. This being so, in the absence of accessory pseudo-brookite, the NRM of basalts can be TRM only in exceptional cases and must normally be CRM.

Kawai et al. (1954) argued for exsolution of ulvöspinel and magnetite on long-term slow cooling of an intermediate quenched titanomagnetite but this is rare in nature and occurs only under highly reducing conditions. From a Curie point analysis of 247 specimens of basalts, Ade-Hall et al. (1965) showed that ulvöspinel exsolution is not a common occurrence while oxidation-induced ilmenite exsolution is.

As in the case of CRM in sediments, it is probable that the CRM's in igneous rocks are induced during geologically short time intervals after solidification, so that the indicated paleomagnetic pole positions are appropriate to the ages of the rocks. However, the effect on paleointensity determinations is more problematical; direct comparison of natural remanence with laboratory induced TRM requires a correction for the relative intensities of CRM/TRM.

Finally, there is another class of rocks and minerals which is finding increasing use in paleomagnetism: non-magnetic materials, e.g., felspars, in which iron oxides are soluble at high temperatures but precipitate as very fine grained magnetic phases at room temperature and so develop CRM. There is no evidence of the rate of precipitation. If it is very slow the CRM may not be contemporaneous with the formation of the host minerals. The particular interest in this type of material arises from the fact that the magnetic grains are very fine, actually single domains in some cases, and have high magnetic stabilities making their paleomagnetic directions particularly reliable.

9.2 THE PROCESS OF CHEMICAL MAGNETIZATION

The acquisition of chemical remanence is physically very similar to the process of thermoremanent magnetization; it is therefore not surprising that the two are hard to distinguish. This applies to remanence in both single domains and multidomains. The simplest case to consider is the formation of independent single domains by the growth of fine grains in a non-magnetic matrix. In this case the only field controlling the magnetization is the ambient external field and there is no grain interaction tending to compromise the single-domain properties. Then for a particular grain volume the probability that the grain moment will change orientation is given by eq.6.1 and the consequent relaxation time for the remanence of an assembly of such grains by eq.6.2. We can now consider the nature of the potential energy barrier in these equations. For hematite grains we suppose that there exists a basal plane anisotropy of the form $K\sin^2 \psi$, due to internal stresses, i.e., the barrier energy for a grain of volume V is VK, so that the relaxation time at temperature T is:

$$\tau = \frac{1}{C}\exp\left(\frac{VK}{kT}\right) \tag{9.1}$$

In magnetite grains, we can treat the barrier energy as being either that required to turn the magnetization from one easy [111] axis to another via a [110] axis, so that by eq.3.16 we can substitute:

$$K = -\left(\frac{K_1}{12} + \frac{K_2}{27}\right) = 1.3 \cdot 10^4 \text{ erg/cm}^3 \tag{9.2}$$

as in eq.6.28, or if shape anisotropy is responsible for the barrier, the energy is that required to turn the magnetization via a small grain axis:

$$K = \frac{1}{2}(N_b - N_a)I_s^2 \tag{9.3}$$

as in eq. 4.12. The effective anisotropy by eq.9.3 is dependent upon grain shape, but may be as large as $1.4 \cdot 10^6$ erg/cm^3 for elongated rod-shaped grains.

During the development of a grain at constant temperature V grows through a critical range in which τ rapidly becomes very large. In effect there is a critical value of V, the blocking volume, analogous to the blocking temperature observed during the cooling of a grain of fixed V, below which an assembly of such grains is superparamagnetic and above which the magnetic moments are stabilized. This process was demonstrated by Kobayashi (1961, 1962) to occur during the precipitation of cobalt-rich grains in Co-Cu alloy. The CRM acquired by an assembly of such grains is then determined by the alignment of grain moments at the blocking volume. For uniaxial grains (the case assumed for hematite with uniaxial magnetic character within the basal plane) the problem is mathematically similar to TRM of single domains and for a randomly oriented assembly leads to the equivalent of (7.5):

$$\frac{M}{M_S} = F'\left(\frac{\mu_B H}{kT}\right) = \int_0^{\pi/2} \tanh(\mu_B H \cos\theta/kT) \cos\theta \sin\theta d\theta \qquad (9.4)$$

where $M_S = fI_S$, the saturation magnetization of the assembly, assuming a volume fraction f of magnetic grains, and $\mu_B = V_B I_S$ is the value of grain moment at the blocking volume V_B. For $\mu_B H/kT < 1$, eq.9.4 reduces to:

$$\frac{M}{M_S} = \frac{1}{3}\frac{\mu_B H}{kT} \qquad (9.5)$$

but for $\mu_B H/kT > 1$, eq.9.4 approaches saturation more sharply than the Langevin equation for a classical paramagnetic:

$$L(a) = \coth(a) - \frac{1}{a} \qquad (9.6)$$

Eq.9.4 and 9.6 are tabulated in Appendixes 1 and 3.

The precise value of V_B depends slightly upon how fast the grains grow; it is related to the effective time τ_B available for blocking of the moment by eq.9.1, which gives:

$$\frac{V_B K}{kT} = \ln(C\tau_B) \qquad (9.7)$$

the numerical value of this expression being 27.6 if $\tau_B = 10^4$ sec, to represent a typical laboratory situation, (with $C = 10^8$ sec^{-1}) but 45 if $\tau = 10^4$ years as it may be in case of natural exsolution. Thus, substituting for μ_B in eq.9.4 we obtain:

$$\frac{M}{M_S} = F'\left\{\frac{I_S H}{K} \ln(C\tau_B)\right\} \qquad (9.8)$$

We now estimate K in terms of the microscopic coercive force H_c, that is the field which must be applied to a grain to switch its moment without the aid of thermal agitation (or when the volume has grown much larger than V_B), by putting (approximately for a random assembly):

$$H_c = \frac{2K}{I_S} \qquad (9.9)$$

whence:

$$\frac{M}{M_S} = F'\left[2\frac{H}{H_c}\ln(C\tau_B)\right] \tag{9.10}$$

Thus the function F' reduces to the linear approximation given by eq.9.5 if:

$$\frac{H}{H_c} < \frac{1}{2\ln(C\tau_B)} \tag{9.11}$$

i.e., $H < \sim H_c/70$. The value of H_c for hematite grains is very variable since it arises from internal stresses; for fine grains exsolving in a matrix with quite different lattice parameters, H_c may be several thousand Oe and CRM is linear in the inducing field H, up to $H \approx 30$ Oe:

$$\frac{M}{M_S} = \frac{2}{3}\frac{H}{H_c}\ln(C\tau_B) \approx 0.01\,H \tag{9.12}$$

For fine single-domain magnetite grains $H_c \approx 250$ Oe (assuming dominance of shape anisotropy) and:

$$\frac{M}{M_S} \approx 0.1\,H \tag{9.13}$$

up to $H \approx 3$ Oe.

These equations lead directly to the deduction that CRM and TRM have comparable but not identical magnititudes for particular single-domain grains. For CRM we may write:

$$\left(\frac{M}{M_S}\right)_{CRM} = \frac{1}{3}\frac{I_S H}{K}\ln(C\tau_B)_{CRM} \tag{9.14}$$

Note that, since we have assumed randomly oriented uniaxial grains, the saturation remanence is $M_{RS} = \frac{1}{2} M_S$. Equivalent expressions for TRM of grains with a single grain size are:

$$\left(\frac{M}{M_S}\right)_{TRM} = \frac{1}{3}\frac{I_{SB}H}{K_B}\ln(C\tau_B)_{TRM} \tag{9.15}$$

or for grains uniformly distributed over a range of sizes up to a definite limit (from eq. 7.18):

$$\left(\frac{M}{M_S}\right)_{TRM} = \frac{2}{9}\frac{I_{SB}H}{K_B}\ln(C\tau_B)_{TRM} \tag{9.16}$$

Comparing (9.14) und (9.15) and assuming that the characteristic times τ_B are comparable in the two cases:

$$\frac{M_{CRM}}{M_{TRM}} \approx \frac{I_S}{I_{SB}} \cdot \frac{K_B}{K} \tag{9.17}$$

for which a typical numerical value is ~ 0.4. The difference is obviously important in deductions about the paleointensity of the geomagnetic field.

If grains grow larger than the critical size for single domains, nucleation of multidomain structure occurs leading to a reduction in magnetization, as was clearly shown by Kobayashi's (1961), 1962) experiment on the growth of Co-Cu grains in a copper matrix. At each stage in the development of a domain structure the local readjustment is sufficiently great that it can be regarded as the formation of a new domain structure, upon which the ambient field will impose a bias. After passing the critical size, the grains will initially be in the pseudo-single domain range in which they still have strong permanent moments, but these will become progressively less significant as the grains grow and the CRM is reduced to a value characteristic of multidomain CRM, as in Fig.9.1. In

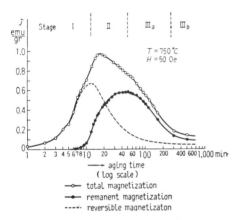

Fig.9.1. Magnetization in a field of 50 Oe of a 2% cobalt-in-copper alloy in which magnetic cobalt-rich grains are growing with time at 750°C. (Figure reproduced by permission from Kobayashi, 1962.) The indicated stages are *I*: growth of grains in the superparamagnetic range showing an increasing susceptibility but no remanence; *II*: the onset of blocking with the establishment of remanence, but diminishing susceptibility; *IIIa*: shows the range in which the grains are pseudo-single domains, intermediate between true single domains and multidomains leading to *IIIb*: the beginning of the true multidomain range.

the multidomain range the magnetization is expected to assume the equilibrium value which is reduced by the factor $1/(1 + N\chi_i)$ when the ambient field is removed, leaving:

$$N_{CRM} = f \frac{H}{N(1 + N\chi_i)} \approx 0.04 fH \tag{9.18}$$

for dispersed magnetite. This differs from the corresponding TRM (eq.7.10) by the factor $I_{SB}/I_S \approx 0.4$. Unfortunately, experiments by both Haigh (1958) and Kobayashi (1959) on the reduction of hematite to magnetite used powders in concentrated form, to which the present equations cannot be applied. Pucher (1969) found the ratio of CRM to TRM in dispersed magnetite to be strongly field dependent, varying from 0.17 to 1.0 as the inducing field was increased from 0.4 to 7.0 Oe.

The chemical remanence induced in a mineral which is precipitated from a pre-existing magnetic mineral is more difficult to treat theoretically. In general we must consider not merely the magnetostatic field exerted by the original mineral on the new one but also the exchange interaction across the phase boundary (Vlasov et al., 1967a). The relative importance of the two constraints depends upon the circumstances and in particular the domain structure of the new mineral; exchange interaction will be dominant in single domains but less important in multidomains. In general the resulting magnetization will have a direction between those of the pre-existing mineral and of the ambient field, although the possibility of developing a reverse magnetization (Chapter 12) must be allowed where exchange interactions are important (Hedley, 1968).

9.3 THE STABILITY OF CHEMICAL REMANENCE

Laboratory tests on the stability of CRM reported by various authors have given divergent results. The generally favoured conclusion is that stabilities of CRM and TRM are similar and the demagnetization curves for a sample of magnetite by Kobayashi (1959) – Fig.9.2 – illustrate this case. However, the processes of oxidation and exsolution of

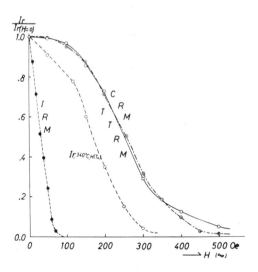

Fig.9.2. Alternating field demagnetization curves of chemical remanence, total thermoremanence and 30 Oe isothermal remanence for a magnetite sample. (Reproduced by permission from Kobayashi, 1959.) The chemical remanence was induced at 340°C and also shown is the demagnetization of a partial thermoremanence induced by cooling in a field from 340°C to room temperature, demonstrating that the CRM is more like the total thermoremanence than partial thermoremanence.

titanomagnetites are more complicated than the reduction of hematite to magnetite. The first stage, a partial oxidation of titanomagnetite in a rock, is accompanied by a reduction in intensity of the original NRM and a relatively unstable CRM. This is replaced during the breakdown of the resulting titanomaghemite to Ti-poor titanomagnetite, which carries a new stable CRM, and ilmenite, which is magnetically passive. Wasilewski (1969) has observed this phenomenon. There have been suggestions that the relative stabilities of CRM and TRM are strongly dependent upon grain size, but these reports probably arise from physical differences after CRM and TRM treatments of the materials studied.

Chapter 10

ALTERNATING FIELD DEMAGNETIZATION AND ANHYSTERETIC MAGNETIZATION

10.1 THE DEMAGNETIZATION METHOD

One of the important techniques in paleomagnetism is the partial demagnetization of rocks by the application of an alternating field slowly reduced to zero amplitude from a selected initial value, usually a few hundred oersteds. During the process the specimens are rotated about two or three mutually perpendicular axes, at rates different from one another and lower than the frequency of the field, so that the specimens are presented to the demagnetizing field in all orientations. This process removes the less stable secondary components of natural remanence, which may be viscous in origin or isothermally induced by fields associated with lightning strikes, leaving most of the primary TRM or CRM which is to be measured. The method, known colloquially as "A.C. washing" or "magnetic cleaning" was established by As and Zijderveld (1958), Creer (1959) and others when the inadequacy of raw ("uncleaned") paleomagnetic observations was becoming apparent.

The rotation of a specimen also minimizes the biassing effect of any stray, steady field. If, during demagnetization of a non-rotating specimen, a small steady field is superimposed upon the alternating field then the decreasing hysteresis loops are biased to one side and after removal of the alternating field the specimen is left with a residual *anhysteretic* magnetization (section *10.3*). Even harmonics in the alternating field have a similar effect, giving different peak values to the opposite polarities of the alternating field, even though the mean value is zero. In demagnetization work stray, steady fields are avoided, in particular by cancellation of the earth's field, and care is taken to use a pure sinusoidal alternating field as nearly as possible.

Repeated measurements of remanence after partial demagnetization in a series of increasing peak fields gives a demagnetization curve, from which one can select the peak field for routine use with a series of similar specimens, usually 150–300 Oe in igneous (titanomagnetite-bearing) rocks but more in hematite-bearing sediments. The field by which any magnetic constituent is demagnetized is characteristic of its magnetic hardness, being comparable to the coercivity of remanence, and each specimen has grains with a range of hardnesses. For thermoremanence or chemical remanence virtually no demagnetization occurs in fields smaller than the coercivity of remanence for the softest grains; the demagnetization curve follows the form shown in Fig.9.2. By choosing a peak field corresponding to the steep part of the curve one chooses to remove the softer half (approximately) of the primary remanence in a specimen. On the other hand isothermal and viscous remanences are removed much more easily, as illustrated by the IRM curve of

Fig.9.2. They are generally characterized by an initial decrease in magnetization with the increasing demagnetizing field. Therefore, in selecting a peak field for routine use with a series of specimens it is important to know that it is beyond the range of this initial decrease in magnetization. It does not normally matter that part of the primary remanence has also been destroyed, so long as all of the remaining remanence is of primary thermal or chemical origin.

A very good illustration of the reduction in scatter of directions of magnetization by alternating field treatment is provided by the data of A. V. Cox, reproduced in Fig. 10.1. Striking examples such as this are most easily explained as the removal of widely scattered secondary magnetizations due to lightning. Secondary magnetizations of vis-

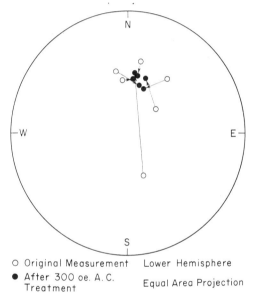

○ Original Measurement Lower Hemisphere
● After 300 oe. A. C. Treatment Equal Area Projection

Fig.10.1. Diminution in scatter of the directions of magnetization of a set of volcanic rocks by 300 Oe alternating field treatment.
(Figure reproduced from Cox and Doell, 1960.)

cous origin are more subtle, being more consistent over large rock bodies. However, their removal is at least as important.

10.2 MAGNETOSTATIC FORCES IN THE DEMAGNETIZATION PROCESS

The magnetization of hematite grains is always so slight that magnetostatic forces arising from the magnetization itself can be neglected in theories of magnetic properties. Such grains can be demagnetized by externally applied fields comparable to their coercive forces because the applied fields act directly on the material of the grains. In magnetite grains this is not so. A demagnetizing field, either steady or alternating, induces a magnetization such that an opposing internal field counteracts the external field. This is apparent in the conclusion of section *4.4* that coercivity of remanence of magnetite grains is of order 5 times the bulk coercivity of the material. Similarly the demagnetizing effectiveness of an applied alternating field is reduced by a factor of about 5.

The process of demagnetization of multidomain grains may be visualized by reference to the expanded picture of the central portion of a hysteresis loop in Fig.10.2. A grain of initial remanence I_R has a self-demagnetizing field $-NI_R$ and before demagnetization it is located at *A* in the figure. It is first subjected to a moderate alternating field,

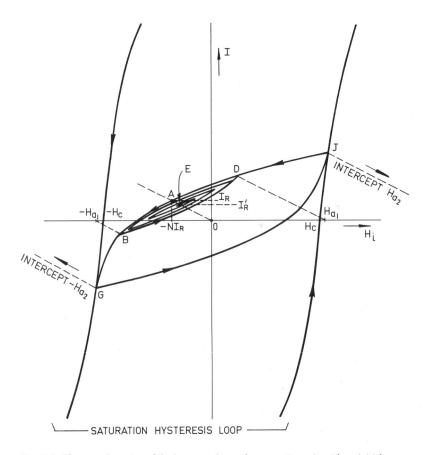

Fig.10.2. The central portion of the hysteresis loop of a magnetite grain with an initial remanence I_R, in the process of its demagnetization by an alternating field. Note that the rectilinear plot gives magnetization I as a function of the internal field H_i and that the external field at any point is found by the Néel construction, as the intercept of one of the broken lines, of gradient $-1/N$, from that point to the field axis.

that is a field of amplitude H_{a1} much less than the coercivity of remanence of the grain, so that its state of magnetization is cycled around the loop $ABCDA$. As the amplitude of the alternating field is gradually reduced so the loop closes and after complete removal of the field the grain is left in the partially demagnetized state E. Note that both the initial and final states, A and E, are on the line of gradient $-1/N$ through the origin.

Now, consider an alternating field of magnitude H_{a2}, just sufficient to demagnetize the grain completely. The magnetic state then oscillates between G and J, which are points on the saturation hysteresis loop (which would be observed if the alternating field had a very large amplitude, $\sim$ 10 kOe). Since the gradients of the limbs of the saturation loop are very steep the internal fields H_i acting on the material of the grain at G and J are little different from the coercive force points H_c, respectively. Thus the amplitude of the externally applied field producing the loop $GJG...$ is approximately:

$$H_{a2} \approx H_c(1 + N\chi_i) \approx H_{CR} \approx 5\,H_c \quad (10.1)$$

Thus it is important that the strength of the demagnetizing field required to destroy the remanence of a sample should not be referred to as its coercivity. It is more nearly correct to refer to it as the coercivity of remanence H_{CR}, although it somewhat larger even than that.

When the grain is subjected to an alternating field sufficient to cause only partial demagnetization, i.e., to move those domain walls which move most easily, the internal field oscillates about values determined by the applied field H_{a1} and the remanence which is left after gradual removal of the field:

$$\left(-\frac{H_{a1}}{1+N\chi_i} - NI'_R\right) < H_i < \left(\frac{H_{a1}}{1+N\chi_i} - NI'_R\right) \quad (10.2)$$

The shape of the demagnetization curve of a material depends upon its properties, such as grain size and coercivity, and also upon the nature of the remanence. But it is a necessary feature of multidomains, even when they have been given a stable remanence such as TRM, that some demagnetization occurs even in quite moderate alternating fields, whereas single domains may completely resist demagnetization until a field comparable to the coercivity of remanence has been applied.

A fortunate consequence of the effective reduction in strength of a demagnetizing field by the internal fields within magnetite grains is that any deflection of thermoremanence away from the direction of the inducing field by the intrinsic anisotropy of a specimen is reduced in the process of partial demagnetization. This effect has been observed by Kern (1961a) whose data are reproduced in Fig.10.3. We still have no detailed quantitative theory of this effect, but qualitatively it follows directly from eq.10.1. If we have a grain with semi-axes $a > b$, then its demagnetizing factors are $N_b > N_a$, so that $(H_i)_a > (H_i)_b$ and demagnetization is more effective in the a direction, in which the initial thermoremanence is disproportionately strong (section 7.8).

It remains to consider how pseudo-single domain thermoremanence responds to

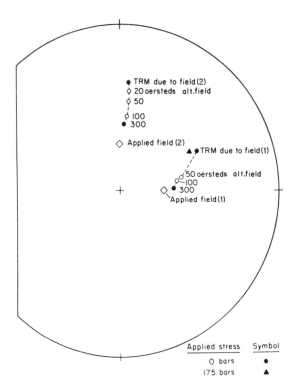

Fig.10.3. Directions of thermoremanent magnetization and inducing fields in a specimen of anisotropic rock with the changes in direction of remanence due to partial demagnetization in a series of increasing alternating fields. The triangle represents a measurement of TRM induced under stress.
(Figure reproduced by permission, from Kern, 1961a.)

alternating field treatment. The fine scale of the domain structure responsible for the p.s.d. effect ensures that it is magnetically hard, but it is for this reason rather more effectively exposed to the demagnetizing influence of an applied alternating field. Consider two p.s.d. moments, represented by A and B in Fig.10.4, in the surface of a grain subjected to a field sufficient to demagnetize the body of the grain but not sufficient to reorient the p.s.d. moments. The magnetization I of the body of the grain includes an irreversible contribution and is therefore stronger than the reversible magnetization of the p.s.d. skin, even without allowing for the lower intrinsic susceptibility of the latter. Inhomogeneity of the grain magnetization then enhances the strength of the field at B. Since specimens are rotated with respect to the field in the demagnetization process, all p.s.d. moments are in turn subjected to a demagnetizing field H_B. Thus although in the process of successively increasing an applied alternating field the p.s.d. moments may be the last to be demagnetized, once the body of the grain is demagnetized further

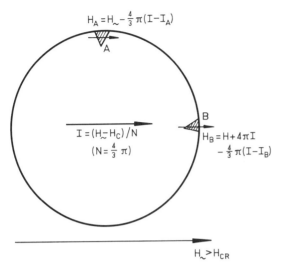

Fig.10.4. Fields experienced by two moments, A and B, in the pseudo-single domain skin of a magnetic grain. An alternating external field is applied and the instantaneous value H exceeds the coercivity of remanence of the bulk of the grain H_{cR}, but is not necessarily sufficient to reorient the harder p.s.d. moments. If the applied field is insufficient to demagnetize the p.s.d. skin then the effective internal fields acting at A and B are as shown. The field acting at B is enhanced by an amount proportional to the difference between bulk magnetization and skin magnetization at B.

increases in the applied field strength become much more effective in demagnetizing the skin. This reduces the "tail" at the high field end of the demagnetization curve.

10.3 ANHYSTERETIC REMANENT MAGNETIZATION (ARM)

If a rock (or laboratory prepared specimen) is subjected to a decreasing alternating field, as in the demagnetization process, but with a small, steady field superimposed, then instead of being demagnetized it is given a remanence approximately proportional in strength to the steady field and known as anhysteretic remanence or ARM. The process provides a useful tool in the study of fundamental magnetic properties in rocks but has been very little exploited.

The equation for the strength of ARM as a function of inducing field in multidomain magnetite grains is derived by reversing the demagnetization argument of section 10.2. We consider a grain whose hysteresis loop is represented in Fig.10.5, the oscillatory external field being biassed by a small, steady field h. Initially an alternating field of strength H_a, substantially greater than the coercivity of remanence of the grain, is applied and the magnetization follows the asymmetrical cycle $OABAB$... By virtue of the superposition of an alternating field of amplitude H_a upon the small steady field h, the total

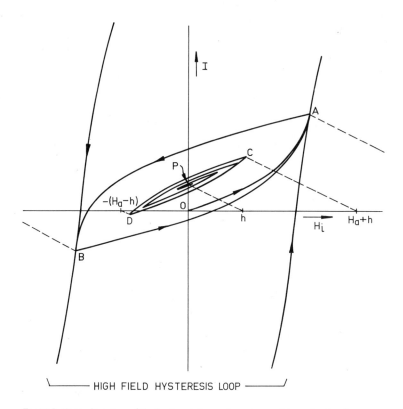

Fig.10.5. Central portion of the hysteresis loop of a grain acquiring anhysteretic remanence in a small, steady field, h. In a strong alternating field the hysteresis loop is $ABAB$ – – – which is slowly reduced through the loop $CDCD$ – – – to converge at the point P at which the internal field acting on the material of the grain is zero.

externally applied field oscillates between values $(H_a + h)$ and $-(H_a - h)$. As the alternating field strength is reduced, so the hysteresis loop reaches a stage represented by the loop $CDCDC...$, whose axis has a gradient which we may equate to the intrinsic susceptibility χ_i, although since we are considering a hysteresis loop, there is in fact no single simple value for χ_i. On this loop the material oscillates symmetrically about the state of zero internal field; the bias h in the external field is precisely compensated by the bias in magnetization I_P at the point P, where $I_P = h/N$. As the hysteresis loop closes to the point P the magnetization converges to the value I_P. In this state the internal field is zero, although the external field is still h.

Now when h is removed, the grain magnetization relaxes along the line PQ in Fig. 10.6. The point Q represents the state of a grain with the measured anhysteretic remanence I_{ARM}, at which the external field has been reduced to zero and the internal demagnetizing field is $(-NI_{ARM})$. It is the intercept of the demagnetization curve PQ with the line OQ of gradient $-1/N$ through the origin. The value of anhysteretic remanence is:

$$I_{ARM} = \frac{h}{N} \cdot \frac{1}{1+N\chi_D} \tag{10.3}$$

where χ_D is the susceptibility of the material in response to the self-demagnetizing field, that is the mean gradient of the curve PQ. It is apparent that this is not constant but

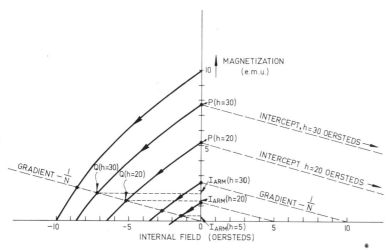

Fig.10.6. Relaxation of magnetization in a magnetite grain with induced ARM as the small steady field h is removed. The final state, Q, is the intercept of the demagnetization curve with the Néel construction, a line of gradient $-1/N$ through the origin. Numerical values are obtained from the data in Fig.10.7.

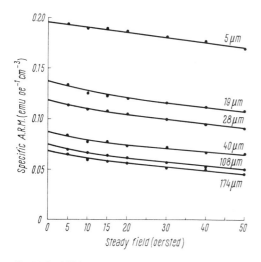

Fig.10.7. ARM per oersted of steady field per unit volume fraction of magnetite grains in six indicated mean sizes as a function of the steady inducing field h.
(Reproduced by permission, from Gillingham and Stacey, 1971.)

increases with the magnitude of I_{ARM} (or h), so that I_{ARM}/h is not constant, but decreases with h, as shown by the data in Fig. 10.7. In fact the data of Fig. 10.7 allow the precise shape of the demagnetization curve to be plotted; since it has the same shape for all values of I_{ARM} the values of I_{ARM} vs. h give the data from which Fig. 10.6 was plotted. The *initial* value of χ_D is only two thirds of the mean alternating field susceptibility which we have referred to as χ_i, but is equal to χ_i for $h \approx 50$ Oe.

If we consider a specimen with a volume fraction $f \ll 1$ of magnetic grains, then its ARM is:

$$M_{ARM} = \frac{fH}{N(1 + N\chi_D)} \approx \frac{fH}{N(1 + N\chi_i)} \tag{10.4}$$

As with TRM (Chapter 7) it is less convenient to compare theoretical and experimental values of M_{ARM} by means of eq. 10.4, which uses the intrinsic susceptibility χ_i, than to use directly values of observed susceptibility χ. Substituting for χ_i by eq. 4.17, eq. 10.4 becomes:

$$M_{ARM} \approx \frac{h}{N} \cdot \frac{\chi}{\chi_i} \approx h\left(\frac{f}{N} - \chi\right) \tag{10.5}$$

Comparison of theory and experiment is made in Table 10.I, which reproduces ARM data on sieved, dispersed magnetite grains by Gillingham and Stacey (1971). Column 4 gives the zero field extrapolations of M_{ARM}/h. Columns 5 and 6 give alternative ways of comparing eq. 10.5 with the observations. Assuming $N = 3.9$ for all grain sizes and $\chi_D = \chi_i$ the calculated ARM is given in column 5, showing that, even with these assumptions, the multidomain theory gives reasonable agreement with observations on the largest grain sizes, but that the observed ARM is significantly stronger than predicted for smaller sizes. This is, of course, immediately suggestive of a pseudo-single domain effect, as with TRM, but we must notice that by eq. 10.5, M_{ARM} is very sensitive to small changes in $1/N$ or χ/f, which are comparable in magnitude. Thus if we calculate the

TABLE 10.I

Values of saturation remanence (M_{RS}), susceptibility (χ) and anhysteretic remanence (I_{ARM}) extrapolated to low inducing fields for dispersed magnetite grains of several mean diameters (d) in terms of volume concentration f, by Gillingham and Stacey (1971) and Gillingham (1971)

d (μm)	M_{RS}/f (e.m.u.)	χ/f (e.m.u.)	M_{ARM}/fh ($h = 0$)	$1/3.9 - \chi/f$	N'
3.4	34.6	0.155	0.196	0.101	2.85
19	17.2	0.207	0.137	0.049	2.91
28	13.6	0.201	0.118	0.055	3.14
40	9.54	0.220	0.087	0.036	3.26
108	6.62	0.198	0.075	0.058	3.66
174	6.16	0.201	0.068	0.055	3.72

values of N required for agreement of the ARM observations with eq. 10.6 (column 6), the range is sufficiently small to be plausible. The case for a pseudo-single domain theory to explain the data is therefore not compelling; the problem will probably be resolved only by measurement on grains of controlled shapes as well as sizes.

Values of saturation remanence, I_{RS} are also given in Table 10.I, because of the possible significance of a linear relationship between $(M_{ARM}/h)^{-1}$ and $(I_{RS})^{-1}$. Combining 10.4 and 4.44, we find:

$$\left(\frac{M_{ARM}}{fh}\right)^{-1} = N\left[1 + (\chi_i H_c)\left(\frac{I_{RS}}{f}\right)^{-1}\right] \tag{10.6}$$

The data in Table 10.I give an acceptable value of N but indicate that $(\chi_i H_c)$ is only half of the value indicated by Parry's (1965) coercivity data in Table 4.I.

A partial anhysteretic remanence (PARM) is produced when a specimen is treated in a decreasing alternating field, but exposed to an additional steady field only over a limited range of the alternating field. Patton and Fitch (1962) showed that, within an experimental uncertainty of about 5%, PARM's acquired over separate alternating field ranges (which together make the total range) and a steady field of 0.43 Oe added in the same way as PTRM's (section 7.7) to equal the total ARM. In more precise measurements up to 50 Oe steady field, D.E.W. Gillingham (1971) found Σ PARM/total ARM to increase linearly with the magnitude of the steady inducing field, up to an average value of about 1.08 at h = 50 Oe. Scatter of the data obscured a possible slight grain-size dependence. The explanation of additivity of PARM's is essentially the same as for PTRM's (section 7.7), but with alternating field agitation acting instead of thermal agitation of the grains.

A final, useful point about ARM is seen by comparing multidomain TRM (eq.7.10) with eq.10.5 for ARM. Since the only factor not common to both is I_S/I_{SB} we obtain immediately:

$$\frac{I_{TRM}}{I_{ARM}} = \frac{I_S}{I_{SB}} \tag{10.7}$$

which is, of course, an average value for a whole specimen. Comparisons of data on specimens which are believed to contain only multi-domained magnetite grains confirm the average value I_S/I_{SB} = 2.5 assumed in Chapter 7. As Dunlop and West (1969) pointed out, eq.10.7 should apply also to single domains. From measurements on four quite different types of specimen with (interacting) single domains, they found values of TRM/ARM in the range 1.6–5.5, closely correlated with values of I_S/I_{SB} estimated from PTRM spectra and spontaneous magnetization curves.

Chapter 11

PIEZOMAGNETIC EFFECTS

11.1 EFFECT OF STRESS ON SUSCEPTIBILITY

As discussed in section 3.2, the application of stress to a magnetostrictive material introduces a magnetic anisotropy of piezomagnetic origin. The theory of the effect as applied to observations in a high field is straightforward because directions of magnetization of all domains are constrained to be close to the field direction and we can observe the angular dependence of magnetic energy, as in Fig.3.7. However, low-field properties are not so simply explained; we have to account for two aspects of the piezomagnetic effect simultaneously. Stress modifies the magnetocrystalline anisotropy and hence the response of domains to an applied field, but this causes rotation of the easy directions of magnetization, so that even in an intrinsically isotropic rock the domain orientations are no longer random. In titanomagnetites, with positive magnetostrictions, domains are rotated away from the axis of a compressive stress. Nagata (1970) has reviewed both the laboratory data on the piezomagnetic effect and its geophysical consequences.

We consider an intrinsically isotropic, titanomagnetite-bearing rock and calculate the variation in susceptibility with stress, in the direction of an axial compression, in terms of the magnetostriction constants λ_{100}, λ_{111}. Then since volume magnetostriction is slight we have a simple way of relating the transverse susceptibility to the longitudinal susceptibility. Appeal is made to the thermodynamic identity between magnetization I and magnetostriction λ in a field H and under stress σ:

$$\left(\frac{\partial \lambda}{\partial H}\right)_\sigma = -\left(\frac{\partial I}{\partial \sigma}\right)_H \tag{11.1}$$

Then since, in low fields, volume magnetostriction is slight, the magnetostriction perpendicular to an applied field, $\lambda_\perp$, is related to the magnetostriction $\lambda_{//}$ in the field direction by:

$$\lambda_\perp \approx -\frac{1}{2}\lambda_{//}$$

Thus, by eq.11.1, the magnetization perpendicular to a stress, $I_\perp$, is related to the parallel magnetization by:

$$\left(\frac{\partial I_\perp}{\partial \sigma}\right)_H = -\frac{1}{2}\left(\frac{\partial I_{//}}{\partial \sigma}\right)_H$$

and hence the stress dependences of susceptibility are related by:

$$\frac{d\chi_\perp}{d\sigma} = -\frac{1}{2}\frac{d\chi_{//}}{d\sigma} \tag{11.2}$$

This relationship is strictly applicable only to susceptibilities in low fields, in which volume magnetostriction is negligible, and under low to moderate stresses, in which induced magnetization strongly outweighs the irreversible (piezoremanent) magnetization. Within these limitations a complicated geometrical calculation of the susceptibility of an assembly of grains having all crystallographic orientations with respect to an applied stress is unnessary. We can calculate all that is required by considering the two simple cases of domains parallel and perpendicular to the stress axis. These are then added in the proportion 1/3 : 2/3 to describe the properties of a random assembly.

Kern (1961b) and Stacey (1962b, 1963) calculated the stress dependence of susceptibility of an assembly of dispersed grains with assumed isotropic magnetostriction λ_s. This assumption gave surprisingly good agreement with observations, but is too crude to serve as a basis for further elaborations of the theory. Also both calculations assumed either explicitly or implicitly that susceptibility was controlled almost entirely by domain rotation effects (or 71° and 109° domain wall movements, which are also stress sensitive) and not significantly by 180° domain wall movements. It is more realistic to suppose that 180° wall movements contribute about 1/3 of the intrinsic susceptibilities of magnetic minerals but that they are unaffected by stress. These objections are avoided in the present approach, which follows closely the theory of Stacey and Johnston (1972).

We consider first a domain oriented along a [111] crystal axis, but then deflected by a small angle θ by the application of a perpendicular field and calculate the magnetostriction in the field direction by eq.3.22. This is then used to give the magnetostrictive strain energy (i.e., change in anisotropy energy) of the domain when a compressive stress σ is applied parallel to the field. We further suppose initially that the geometry is as represented in Fig.11.1, in which the field and the rotation of magnetization are confined to the shaded {110} plane. The result can then be very simply generalized to a random distribution of crystal axes about the {111} magnetization axis by an appeal to symmetry.

The direction cosines α_1, β_1 (eq.3.22) are referred to the x axis (as marked in Fig. 11.1). Then we have:

$$\alpha_1 = \cos\left(\cos^{-1}\frac{1}{\sqrt{3}} - \theta\right) = \frac{1}{\sqrt{3}}(\cos\theta + \sqrt{2}\sin\theta)$$

$$\alpha_2 = \alpha_3 = \left[\frac{1}{2}(1-\alpha_1^2)\right]^{1/2} = \frac{1}{\sqrt{3}}\left(\cos\theta - \frac{1}{\sqrt{2}}\sin\theta\right)$$

$$\beta_1 = \cos\left(\frac{\pi}{2} - \cos^{-1}\frac{1}{\sqrt{3}}\right) = \sqrt{\frac{2}{3}}$$

$$\beta_2 = \beta_3 = \left[\frac{1}{2}(1-\beta_1^2)\right]^{1/2} = -\frac{1}{\sqrt{6}}$$

so that the magnetostriction (in the direction marked λ in Fig.11.1) is:

$$\lambda = \lambda_{100}\left(\frac{1}{4}\sin^2\theta + \frac{1}{\sqrt{2}}\sin\theta\cos\theta\right) + \lambda_{111}\left(-\frac{1}{2} + \frac{5}{4}\sin^2\theta - \frac{1}{\sqrt{2}}\sin\theta\cos\theta\right) \quad (11.3)$$

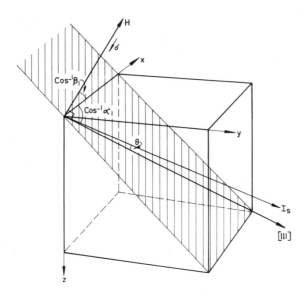

Fig.11.1. Angles used in the calculation of magnetostriction normal to a [111] axis in magnetite, due to the rotation of magnetization by a small angle θ away from the [111] axis. All directions are here confined to a particular (110) plane (shaded) to simplify the calculations, which are then generalized to a random assembly.

If we rotate the crystal structure by 180° about the [111] axis then all odd powers of $\sin \theta$ appear with reversed signs. In averaging over all crystal orientations the odd terms thus disappear and for a random assembly eq.11.3 reduces to:

$$\lambda = -\frac{1}{2}\lambda_{111} + \left(\frac{1}{4}\lambda_{100} + \frac{5}{4}\lambda_{111}\right)\sin^2 \theta \tag{11.4}$$

The magnetostrictive strain energy by virtue of magnetization at θ against an applied stress σ is therefore:

$$E_\lambda = \left[-\frac{1}{2}\lambda_{111} + \left(\frac{1}{4}\lambda_{100} + \frac{5}{4}\lambda_{111}\right)\sin^2 \theta\right]\sigma \tag{11.5}$$

E_λ is added to the anisotropy energy which results from the deflection of magnetization away from the [111] axis, as in eq.3.19:

$$E_K = \left(\frac{K_1}{3} + \frac{K_2}{27}\right) - \left(\frac{2K_1}{3} + \frac{2K_2}{9}\right)\sin^2 \theta + \ldots \tag{11.6}$$

The susceptibility of the grain can now be generalized to account for the effect of σ to obtain $\chi(\sigma)$, as in section 4.2 but with E_λ added to E_K, giving:

$$\chi_\perp(\sigma) = \frac{I_s^2}{-\frac{4}{3}\left(K_1 + \frac{1}{3}K_2\right) + \sigma\left(\frac{1}{2}\lambda_{100} + \frac{5}{2}\lambda_{111}\right)} \tag{11.7}$$

Now if we suppose that we have an assembly of similarly oriented domains in grains of demagnetizing factor N, so that eq.11.7 gives the intrinsic susceptibility in the direction of the field, then the observed susceptibility $\chi(\sigma)$ is related to the intrinsic susceptibility by eq.4.17, and if the magnetic grains occupy a volume fraction f of a specimen:

$$\chi(\sigma) = \frac{fI_s^2}{-\frac{4}{3}\left(K_1 + \frac{1}{3}K_2\right) + \sigma\left(\frac{1}{2}\lambda_{100} + \frac{5}{2}\lambda_{111}\right) + NI_s^2} = \frac{\chi_0}{1+s'\sigma} \quad (11.8)$$

where:

$$s' = \frac{\frac{1}{2}\lambda_{100} + \frac{5}{2}\lambda_{111}}{-\frac{4}{3}\left(K_1 + \frac{1}{3}K_2\right) + NI_s^2}$$

So far we have considered the susceptibility of an assembly of grains in the direction of an applied stress when all of the constituent domains are normal to the stress. In an isotropic rock the domains are distributed equally in all directions, so that the intrinsic susceptibility is not equal to $\chi_\perp$ but is:

$$\chi_i = \frac{2}{3}\chi_\perp + \frac{1}{3}\chi_{111}$$

in which only $\chi_\perp$ is stress-sensitive. Thus to a good approximation the stress-sensitivity of susceptibility of a magnetite- or titanomagnetite-bearing rock is:

$$s_\chi = \frac{2}{3}s' = \frac{\lambda_{100} + 5\lambda_{111}}{3NI_s^2 - 4\left(K_1 + \frac{1}{3}K_2\right)} \quad (11.9)$$

Values for each of the five constants are listed in Table 3.I for a range of compositions of the $(1-x)$ Fe$_3$O$_4 \cdot x$ Fe$_2$TiO$_4$ solid solution series. The corresponding calculated values of s_χ are given in Table 11.I; coincidence with the range observed by Kapitsa (1955) (0.8–3.3 · 10^{-4} cm^2/kg) is convincing. The trend to stronger stress-dependence with increasing Fe$_2$TiO$_4$ content is apparent in the data of Ohnaka and Kinoshita (1968b), although absolute values are not directly comparable because they prepared synthetic powders which were not dispersed in a magnetically inert matrix. The stress dependence of susceptibility, χ_p, in the direction of a compressive stress σ, over a wide stress range is well represented by an equation of the form 11.8:

$$\frac{\chi_p(\sigma)}{\chi(o)} = \frac{1}{1+s_\chi\sigma} \quad (11.10)$$

and the stress dependence at low stresses is thus:

$$\frac{1}{\chi_p}\frac{d\chi_p}{d\sigma} = -s_\chi \quad (11.11)$$

TABLE 11.I

Piezomagnetic properties of titanomagnetite-bearing rocks for a series of titanomagnetite compositions $(1-x)Fe_3O_4 \cdot xFe_2TiO_4$, estimated from the data in table 3.1 by Stacey and Johnston (1972)

x	S_χ (cm²/kg)	S_R (cm²/kg)	σ_c (kg/cm²)
0	$1.11 \cdot 10^{-4}$	$0.76 \cdot 10^{-4}$	330
0.04	$1.28 \cdot 10^{-4}$	$1.12 \cdot 10^{-4}$	390
0.10	$1.5 \cdot 10^{-4}$	$1.3 \cdot 10^{-4}$	470
0.18	$2.4 \cdot 10^{-4}$	$3.2 \cdot 10^{-4}$	300
0.31	$3.2 \cdot 10^{-4}$	$3.9 \cdot 10^{-4}$	290
0.56	$11.5 \cdot 10^{-4}$	$18.7 \cdot 10^{-4}$	130

S_χ is the stress sensitivity of susceptibility by eq.11.9, S_R the stress sensitivity of remanence by eq.11.22 and σ_c the critical stress for major domain readjustment by eq.11.26.

From the symmetry argument, with negligible volume magnetostriction, it follows that the stress dependence of susceptibility χ_n normal to a compressive stress is:

$$\frac{1}{\chi_n}\frac{d\chi_n}{d\sigma} = +\frac{S_\chi}{2} \tag{11.12}$$

The general equation for χ_n at arbitrarily high stress is not obvious because at high stresses domain alignment becomes substantial and the contribution of $\chi_\parallel$ to χ_n is increasingly important. However, for most problems, including the significance of magnetostriction in paleomagnetism and tectonomagnetism, it is sufficient to assume $S\sigma \ll 1$, so that eq. 11.11 and 11.12 apply.

11.2 EFFECTS OF STRESS UPON REMANENCE

The modification of magnetocrystalline anisotropy by stress and the consequent deflections of easy directions of magnetization in a crystal necessarily imply that the domains and therefore remanence are deflected in the same way. In principle, calculation of the deflection of remanence in an isotropic rock requires an integration over random crystallographic and domain orientations, but the problem can be simplified by appeal to symmetry, as in the susceptibility problem considered in section *11.1*. Kern (1961b) approached the problem by supposing that eq.11.1 is applicable to remanence, as it is to induced magnetization and thus obtained the same equation for the variation of both with stress. His calculation was justified by the fact that the stress dependences of susceptibility and remanence are quite similar. However, the appeal to eq.11.1 assumed thermodynamic reversibility of the magnetization, which is satisfactory for induced magnetization but is inappropriate for remanence. Johnston and Stacey (1972) therefore avoided the implication of thermodynamic reversibility by treating the remanence of an assembly of grains as a sum of individual domain moments and not as a quantity which can be treated directly in terms of the equations describing piezomagnetism. Their approach is followed here.

We consider an assembly of grains with a net remanent magnetization and apply a stress in the direction of the magnetization. Then those domains oriented within an elementary cone of semi-angle ϕ and thickness $d\phi$ are deflected by the stress to larger semi-angles $(\phi + \theta)$ (Fig.11.2). The angles of deflection θ are functions not only of the stress

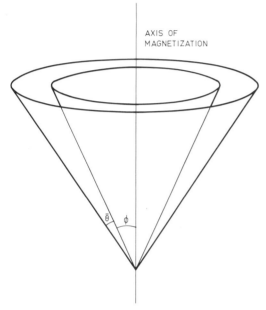

Fig.11.2. Domains oriented within an elementary cone of semi-angle ϕ about the axis of magnetization of an assembly are deflected by stress to a wider cone of average semi-angle $(\phi + \theta)$, where θ is represented by eq.11.16.

σ and the angle ϕ, but also of the orientation of each crystal with respect to the stress axis. Since (for titanomagnetite) the domain magnetization (unstressed) is aligned with a [111] axis, the crystal orientation can be represented by the azimuthal angle of the stress axis with respect to a (110) plane containing the magnetization direction. Determination of the average deflection, $\bar{\theta}$, thus requires an integration over all azimuthal angles. However, cubic crystal symmetry ensures that values of anisotropy and magnetostrictive strain energies, E_K, E_σ, repeat themselves at $120°$ intervals of the azimuthal angle and by averaging the two values obtained by considering stress axes within the (110) plane, i.e., at azimuthal angles $0, \pi$ (Fig.11.3) we obtain the average values of E_K, E_σ for all azimuthal angles:

$$\bar{E}_K = \frac{1}{3}(K_1 + \frac{1}{3}K_2)(1 - 2\sin^2\bar{\theta}) \tag{11.13}$$

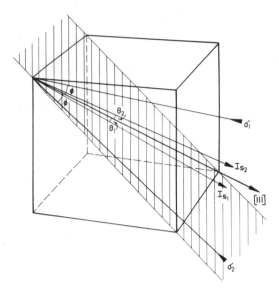

Fig.11.3. Stress at an angle ϕ to the [111] axis of a domain magnetization causes a deflection θ which is a function of the azimuthal angle of the stress axis to the shaded (110) plane. By symmetry the average value of θ is the average $\frac{1}{2}(\theta_1 + \theta_2)$ of the deflections due to stresses within the (110) plane in the directions indicated.

$$E_\lambda = \bar{\lambda}\sigma = -\sigma\left(\lambda_{100} + \frac{1}{2}\lambda_{111}\right)\sin 2\phi \sin\bar{\theta} \qquad (11.14)$$

where $\bar{\lambda}$ is the mean magnetostriction in the direction of σ at ϕ to [111]. Then since θ is determined by the condition of minimum total energy:

$$d(E_K + E_\lambda)/d\theta = 0 \qquad (11.15)$$

and we obtain the mean deflection:

$$\bar{\theta} = \frac{\sigma\left(\lambda_{100} + \frac{1}{2}\lambda_{111}\right)}{-\frac{4}{3}\left(K_1 + \frac{1}{3}K_2\right)} \sin 2\phi \qquad (11.16)$$

As is intuitively reasonable, the deflection is greatest at $\phi = \pi/4$ when the stress is aligned at 45° to a domain.

We can now examine the contribution to the total remanent magnetization of the collection of grains by those which have orientations parallel or antiparallel to the elementary cone of vectors in the range ϕ to $(\phi + d\phi)$ in the absence of stress. They represent a volume fraction:

$$\frac{dV}{V} = \sin\phi \, d\phi \qquad (11.17)$$

EFFECTS OF STRESS UPON REMANENCE

There is a net alignment of these domains in the direction of the axis, which is presumed to have been caused by an earlier application of an axial field, whose aligning influence was proportional to $\cos \phi$. Then if the magnetization process was linear, as in the case of thermoremanence (Chapter 7) or anhysteretic remanence (Chapter 10), the magnetization at angle ϕ due to the elementary cone of domains is:

$$dI = A \cos \phi \cdot V \sin \phi \, d\phi \tag{11.18}$$

where A is a constant which may be substituted from the appropriate equation for the magnetization process. When the domains are deflected by stress their contribution to the axial magnetization of the assembly is:

$$dI_{//} = AV \cos \phi \sin \phi \cos(\phi + \bar{\theta}) \, d\phi \tag{11.19}$$

Since we have restricted our interest to small deflection angles the expansion of $\cos(\phi + \bar{\theta})$ is simplified by putting $\sin \bar{\theta} = \bar{\theta}$ and $\cos \bar{\theta} = 1$, which allows a direct substitution for $\bar{\theta}$ by eq.11.16:

$$dI_{//} = AV(\cos^2 \phi \sin \phi - 2C \cos^2 \phi \sin^3 \phi) d\phi \tag{11.20}$$

where:

$$C = \frac{\sigma\left(\lambda_{100} + \frac{1}{2}\lambda_{111}\right)}{-\frac{4}{3}\left(K_1 + \frac{1}{3}K_2\right)}$$

Then integrating over the whole assembly, i.e. $\phi = 0 \to \pi/2$:

$$I_{//} = \frac{AV}{3}\left(1 - \frac{4}{5}C\right) \quad \text{or} \quad I_{//}(\sigma) = I_0(1 - s_R \sigma) \tag{11.21}$$

where:

$$s_R = \frac{3}{5} \cdot \frac{\lambda_{100} + \frac{1}{2}\lambda_{111}}{-\left(K_1 + \frac{1}{3}K_2\right)} \tag{11.22}$$

This is defined as the stress sensitivity of remanence and numerical values are given in Table 11.I. Similarly, if compressive stress is applied in a direction perpendicular to the remanence:

$$I(\sigma) = I_0\left(1 + \frac{1}{2}s_R \sigma\right) \tag{11.23}$$

If a high stress is applied then the linear approximations represented by (11.21) and (11.23) are no longer adequate. Stress may then also have an irreversible effect upon remanence. But, as shown by Ohnaka and Kinoshita (1968a) irreversible effects of stress upon TRM in stable rocks are significant only under stresses exceeding 1 kbar so that (11.21) and (11.23) suffice for consideration of paleomagnetic implications of magnetostriction (section *11.3*) and for the role of remanence in seismomagnetic and volcanomagnetic effects (section *11.4*).

Nagata and Kinoshita (1967) reported that the application of hydrostatic pressure to magnetite increased its magnetostriction by 15 % per kbar and decreased the anisotropy constant K_1 by 5 % per kbar. If we take the Curie point isotherm to be at about 20 km depth in the earth's crust then pressures up to 6 kbar are imposed upon magnetite grains in the range of depths of interest in calculations of the seismomagnetic effect (Stacey, 1964; Shamsi and Stacey, 1969), implying an increase up to about 125% in the stress sensitivities of magnetization. However, temperature has a stronger effect upon the constants than does pressure and as seismomagnetic calculations are necessarily very rough approximations to real geological situations nothing is gained by allowing for the pressure effect. What is more important is that the stress sensitivities of induced and remanent magnetizations are very similar (Table 11. I) and do not need to be accounted for separately in seismomagnetic calculations.

Irreversible effects of stress on remanences are generally referred to a piezoremanent magnetization (PRM). This has received much attention by T. Nagata, H. Kinoshita and co-workers, whose results are conveniently summarized by Fuller (1970). PRM due to moderate stresses is akin to IRM in strength and stability, so that in low fields, comparable to that of the earth, it is not significant, except perhaps for rocks with titanomagnetite very rich in Fe_2TiO_4, whose stress sensitivities are extremely high (Table 11.I). The essential feature is that the application of stress causes a specimen to approach more nearly its equilibrium magnetization in the ambient field to which it is exposed. Thus rocks acquire PRM if subjected to a stress cycle while exposed to a field but may be partially demagnetized if stressed in zero field.

The IRM which a specimen would acquire in a particular field is roughly doubled as PRM if it is temporarily subjected to a 1 kbar stress during the application of the field. Thus induction of significant PRM requires the application of fields of tens of oersteds. Its variation with intensity of stress is non-linear, the initial effect of stress being proportionately more effective. This observation requires an explanation in terms of the superposition of externally applied stress and internal stress in magnetic grains. If there were no internal stresses then we would expect all piezomagnetic effects to be reversible until some threshold stress was exceeded, when a new domain structure would be set in. However, the presence of localized stresses allows the applied stress to exceed the threshold locally by superposition and so produce small local, irreversible changes in domain structure.

PRM due to stresses up to a few hundred bars can involve only those domain readjustments which are favoured by internal stresses and are therefore unstable in the paleomagnetic sense. Being of low stability it is readily destroyed by conventional alternating field demagnetization techniques. It may contribute to the secondary components of natural magnetizations of rocks (and will certainly do so if they have been subjected to kilobar stresses), but is not a major problem in paleomagnetism.

The magnitude of the threshold stress required for major domain rearrangement may be estimated approximately by considering the energy barrier ΔE_K between [111] axes (via [110]) to be cancelled by the magnetoelastic energy gained when a stress σ parallel to [111] deflects the magnetization to [110]. (Strictly it is the maximum potential gradient which must be cancelled not the potential peak, but, with the particular geometry assumed here, this makes little difference.) The anisotropy energy difference by eq. 3.16 is:

$$\Delta E_K = -\left(\frac{K_1}{12} + \frac{K_2}{27}\right) \quad (11.24)$$

and the change in magnetoelastic energy is:

$$\Delta E_\lambda = -\frac{1}{2}\lambda_{111}\sigma \quad (11.25)$$

Thus the estimated critical stress is:

$$\sigma_c = \left(\frac{K_1}{6} + \frac{2}{27}K_2\right)/\lambda_{111} \quad (11.26)$$

Estimated values of σ_c for rocks with different titanomagnetites are given in Table 11.I. For all compositions up to about 30% Fe_2TiO_4 they fall in the range 300 to 500 bar.

11.3 MAGNETOSTRICTION AND PALEOMAGNETISM

The existence of a piezomagnetic effect in magnetic minerals has been recognized for many years (Wilson, 1922), but interest in it intensified dramatically in the late 1950's when Graham (1956) drew attention to its possible implications in paleomagnetism. At that time many geophysicists disputed the plausibility of continental drift, although in terms of paleomagnetic observations it appeared unavoidable. Thus the possibility that paleomagnetism had a basic flaw, viz., the deflection of natural remanence in rocks by stress, was quickly taken up.

The essential question to be answered was whether the thermoremanence which is induced in an igneous rock sample while it is under stress, as it may be while acquiring natural remanence, is found to depart from the direction of the inducing field when the stress is released, as it would be by the removal of the sample to a paleomagnetic laboratory for measurement. Systematic experiments by Stott and Stacey (1959, 1960) and Kern (1961a) showed that no measurable deviation occurred for any of the isotropic rocks examined, as is indicated by the data reproduced in Table 11.II, in which all of the specimens except the last group were intrinsically isotropic. The serpentine was highly

TABLE 11.II

Mean directions of magnetization of specimens which were stressed during the similar control specimens[1].

Group	Type of rock	No. of Specimens	Load applied
1	various igneous rocks (preliminary survey, Stott and Stacey, 1959)	20	0
		20	500 kg/cm^2
2	Cainozoic basalts, Victoria (sites NV10, NV15, of Green and Irving, 1958)	20	0
		20	500 kg/cm^2
3	Tasmanian dolerite, bore core 5001	13	0
		13	1000 kg/cm^2
4	Silurian porphyry, Yass, N.S.W.	5	0
		5	500 kg/cm^2
5	Tasmanian dolerite, Melville Creek, Red Hill, Tasmania	4	0
		4	500 kg/cm^2
6	Tasmanian dolerite, bore core 5001, heated specimens from group 3	6	800 kg/cm^2
7	Tholeiitic basalt, Berrima, N.S.W. (site B, Green and Irving, 1958)	6	0
		6	500 kg/cm^2
8	Cainozoic basalts, Victoria, heated specimens from group 2	4	500 kg/cm^2
9	Cainozoic basalt, Victoria (i)	1	500 kg/cm^2
	(ii)	2	500 kg/cm^2
	(iii)	1	500 kg/cm^2
	(iv)	1	0
10	Serpentine, Tumut Ponds, N.S.W.	4	0
		4	500 kg/cm^2

[1] From Stott and Stacey (1960).

MAGNETOSTRICTION AND PALEOMAGNETISM 157

acquisition of TRM or IRM, compared with directions in unstressed but otherwise

Magnetization	Angle of field to axis of load, deg.	Mean angle of magnetization, deg.	Standard deviation of individual angles from the mean, deg.			
TRM	66	65.2	2.5			
TRM	66	65.4	2.6			
TRM	46	46.2	2.6			
TRM	46	47.0	3.5			
TRM	46	46.5	2.9			
TRM	46	45.5	2.5			
TRM	46	47.6	2.2			
TRM	46	45.8	3.1			
IRM	49	49.8	0.5			
IRM	49	48.8	1.9			
IRM	49	50.0	1.3			
IRM	49	49.8	0.4			
IRM	49	47.3	2.2			
IRM	49	49.5	5.9			
			individual directions			
TRM in nitrogen	46		44			
TRM in air	46		42	44		
TRM in air (loaded before heating)	46		42			
TRM in nitrogen	46		46			
TRM	66		68	63	68	58
TRM	66		10	70	58	75

anisotropic and showed that in this case large deflections are possible, although the specimens were not oriented with respect to their anisotropies so that systematic trends are not apparent. Since only isotropic rocks are of interest in paleomagnetism, these results and a similar set by Kern (1961a) negated the argument that the magnetostrictive properties of rocks invalidated the paleomagnetic method.

The considerations in the previous sections show that an intrinsically isotropic rock becomes anisotropic when stressed and in this condition it must acquire a thermoremanence which is deflected from the field direction. The experiments of Stott and Stacey (1959, 1960) and Kern (1961a) therefore show that the deflection has just the magnitude required to bring the remanence back into the field direction when the stress is released. This observation has an interesting, but not unexpected, implication concerning the nature of thermoremanence. As a magnetic mineral cools, so the values of its magnetic anisotropy and magnetostrictive constants, as well as spontaneous magnetization, vary. This means that the anisotropy due to a constant stress and also the piezomagnetic deflection of remanence vary during the cooling. It follows that the remanence has at each temperature the direction corresponding to the anisotropy at that temperature, so that removal of the stress at any temperature cancels the deflection due to the piezomagnetic anisotropy at that temperature.

It remains to consider the effect of stress applied during chemical or phase changes of magnetic minerals. We are not here concerned with gross metamorphism of a rock, in which stress would produce an obviously anisotropic rock of no use for paleomagnetism, but with relatively slight changes such as the exsolution of hematite and ilmenite from a solid solution during slow cooling. Stott and Stacey (1960) sought such an effect in a sample of basalt (group 9 specimens of Table 11.II) which was oxidized by heating and cooling in air during the acquisition of TRM (plus some CRM). No significant deflection of remanence resulted. Although a single result cannot be regarded as conclusive, it is consistent with the observations on TRM, since if a mineral is formed below its Curie point, but under stress, it is formed with a piezomagnetic anisotropy which deflects the CRM induced in it in the same way as TRM would be deflected. Further cooling leaves it with the anisotropy and remanence deflection corresponding precisely to its room temperature properties, so that if no intrinsic anisotropy is produced, remanence is found to be undeflected when stress is released.

11.4 THE SEISMOMAGNETIC AND VOLCANOMAGNETIC EFFECTS

The search for a geomagnetic effect associated with earthquakes has a long and unsuccessful history. Since about 1960 a satisfactory theoretical basis for it has emerged and instruments of sufficient sensitivity and stability have become generally available, but the seismomagnetic effect is certainly small and difficult to distinguish from background disturbances as well as being highly localised, making planned observations difficult. The stimulus to continued effort is the relevance of the effect to the elusive problem of earthquake prediction, which will almost certainly be the subject of increasing activity in the next decade or so. The most encouraging observations so far are not of the seismomagnetic effect itself, but of the closely related volcano-magnetic effect.

The basis of both is the piezomagnetic effect, according to which the magnetizations of crustal rocks are changed by tectonic stress, causing small, local magnetic anomalies which accompany the changing stress pattern. If the stress pattern at any instant and the magnetic properties of crustal rocks were both known in detail then the form and intensity of the corresponding magnetic anomaly could be unambiguously calculated. However, this is far from being the case and we are reduced to making plausible guesses about both tectonic stress patterns and magnetic properties. Calculations of the seismomagnetic effect are therefore uncertain to the extent that the earthquake mechanism is not understood and the magnetic minerals of the crust are unknown.

A rough estimate of the magnitude of a seismomagnetic anomaly due to a surface earthquake may be made by considering the stress build up (or release) to occur in a spherical volume of rock immediately below the surface, in which the stress sensitivity of the magnetization I is s, so that the piezomagnetic increment in magnetization is:

$$\Delta I = sI\sigma \tag{11.27}$$

The corresponding peak magnetic field increment at the surface is then:

$$\Delta H \approx \frac{4}{3}\pi\Delta I \tag{11.28}$$

Assuming $s = 2 \cdot 10^{-4}$ cm^2/kg, $I = 10^{-3}$ e.m.u. and $\sigma = 50$ kg/cm^2, we obtain $\Delta H \approx 4 \cdot 10^{-5}$ Oe = 4 gammas. This is a reasonably conservative estimate as the magnetization may be 10^{-2} e.m.u. or more in a very favourable area. The stress is unlikely to be much higher; the estimate for the San Francisco 1906 shock is 75 kg/cm^2, but for many large shocks it is lower. Thus we can do little better than make an order-of-magnitude estimate that seismomagnetic effects have local intensities which are characteristically a few gammas, but may in particularly favourable cases be several tens of gammas and in unfavourable cases only a fraction of a gamma (and therefore virtually unobservable against background noise). More detailed calculations require explicit account to be taken of the seismic stress pattern.

The stress at any point (x, y, z) in the crust may be resolved into the three principal normal stresses, $\sigma_1, \sigma_2, \sigma_3$ (assumed positive for compression), so that the magnetization I of the unstressed rock at that point can also be resolved into components I_1, I_2, I_3 in the directions of the principal stresses. Then in terms of the stress-sensitivity of magnetization (s_χ for induced magnetization, by eq.11.9, or s_R for remanence, by eq.11.22) these components of magnetization are changed by amounts:

$$\Delta I_1 = s I_1 \left(-\sigma_1 + \frac{\sigma_2}{2} + \frac{\sigma_3}{2}\right)$$

$$\Delta I_2 = s I_2 \left(\frac{\sigma_1}{2} - \sigma_2 + \frac{\sigma_3}{2}\right) \quad (11.29)$$

$$\Delta I_3 = s I_3 \left(\frac{\sigma_1}{2} + \frac{\sigma_2}{2} - \sigma_3\right)$$

These are the components of the incremental moment due to the piezomagnetic effect. The contribution of such an incremental moment in a volume (dx, dy, dz) at (x, y, z) to the piezomagnetic anomaly at a point on the surface (x', y', z') is most simply calculated by resolving the moment into x, y, z components, $\Delta I_x, \Delta I_y, \Delta I_z$ and applying the dipole law of force, which gives the incremental field at the surface. Integrating over all volume elements we obtain the x, y, z components of the anomaly field:

$$F_x = \int_{-\infty}^{\infty}\int_{-\infty}^{\infty}\int_{-\infty}^{\infty} \frac{[\Delta I_x(2x_1^2 - y_1^2 - z_1^2) + \Delta I_y(3x_1 y_1) + \Delta I_z(3x_1 z_1)]}{(x_1^2 + y_1^2 + z_1^2)^{5/2}} dx\,dy\,dz \quad (11.30)$$

and similar expressions for F_y, F_z with x_1, y_1, z_1 appropriately interchanged, where

$$x_1 = x' - x; \quad y_1 = y' - y; \quad z_1 = z' - z.$$

All calculations to date have assumed uniform rock magnetization from the surface to the Curie point isotherm at about 20 km (below which it becomes zero) and $s = s_R \approx 10^{-4}$ cm^2/kg (which is seen from the data in Table 11.I to be a conservative estimate) with straightforward stress patterns represented by analytical expressions. Probably closest to reality were the piezomagnetic calculations applied to mathematical models of the San Francisco (1906) and Alaska (1964) earthquakes by Shamsi and Stacey (1969). The work of Breiner and Kovach (1967) with an array of rubidium vapour magnetometers along the San Andreas fault in California indicates the technical difficulties involved in checking these predictions, but they nevertheless achieved significant success in recording very small magnetic precursors to creep (aseismic) movements of the fault.

Similar piezomagnetic calculations were made by Stacey et al. (1965) for stresses associated with eruptions of Caribbean volcanos and the essential validity at least of the principle has been strikingly demonstrated by observations of the volcanomagnetic effect on two New Zealand volcanos (Johnston and Stacey, 1969a,b). Fig.11.4 shows the progressive change in the difference between the magnetic field strengths at two sensors over a few days before an eruption of Mt. Ruapehu on the 5th of April, 1968.

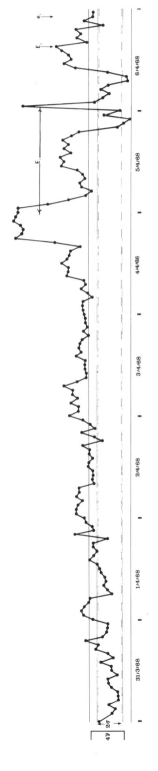

Fig.11.4. Hourly mean values of the difference in magnetic field strength of two sites, 8 km apart, near to Mt. Ruapehu, New Zealand, showing a progressive trend before the eruption on 5 April 1968. The maximum anomaly field is about 10 gammas. Horizontal lines indicate the standard deviation range for volcanically quiet times. They bracket most of the measurements except during times of significant magnetospheric disturbance. (From Johnston and Stacey, 1969a.)

Chapter 12

REVERSALS OF REMANENT MAGNETIZATION

12.1 EVIDENCE FOR REVERSALS OF THE GEOMAGNETIC FIELD

The discovery of reversed polarity remanent magnetization in rocks is about 100 years old and various people have been credited with the discovery. Smith (1971) pointed out that the earliest recognition of reversed NRM appeared to be J. A. Broun's field-observation in southern India in 1855 but Broun did not attribute his discovery to an ancient geomagnetic field-reversal. In 1899, G. Folgheraiter, and in 1905, B. Brunhes noted that certain rocks indicated a once inverted magnetic inclination in Europe, but M. Matuyama and P. L. Mercanton in the 1920's appear to have been first to consider the rapid field-reversals which are now recognized to have occurred many times. But even in the 1950's the reality of field-reversal was not generally proven. Néel (1951) had questioned the absence of reversed NRM in a Jurassic to Miocene sedimentary sequence in Colorado, studied by Torreson et al. (1949), while lavas of similar age had been found elsewhere with reversed NRM. Even as late as 1962, while recognizing that the evidence of field reversals was very strong, Blackett (1962) suggested that it was not possible to exclude completely the possibility that all the reversed rocks had acquired their opposite polarities by unknown physico-chemical mechanisms in their magnetic minerals. This is the hypothesis of self-reversal as opposed to field-reversal; the mechanisms for self-reversal are discussed in section *12.2*. The particular discovery of laboratory self-reversal in Haruna dacite (Nagata, 1952) established the reasonableness of the self-reversal hypothesis and so delayed by many years general acceptance of field-reversals.

With the accumulation of paleomagnetic data, the approximate equality of numbers of "normal" and reversely magnetized rocks became established and this is curcumstantial evidence for magnetization by a geomagnetic field which changes polarity and may have either polarity with equal probabilities. But without precise dating and correlation between measurements on independent rocks, even massive numbers of polarity data are inconclusive. More definite evidence of field reversals was obtained by Cox et al. (1963) and McDougall and Tarling (1963) from the comparison of polarities as a function of potassium-argon ages of rocks from different parts of the world. Fig.12.1 reproduces a recent compilation with a time-table of reversals by Cox (1969), including an indication of the number of data upon which the table is based. It is quite implausible that self-reversal mechanisms should begin and end synchronously in different rock types and in different parts of the earth. However, since the possibility of self-reversal is well recognized, magnetic polarities of individual rocks are treated with caution.

EVIDENCE FOR REVERSALS OF GEOMAGNETIC FIELD 163

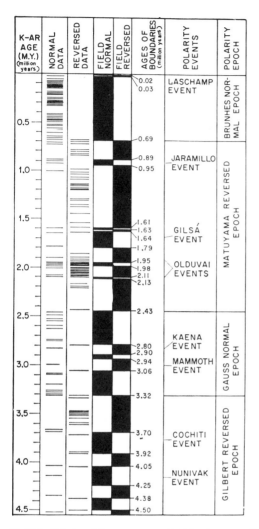

Fig.12.1. Polarity of the geomagnetic field for the past million years.
(Reproduced by permission, from Cox, 1969.)

Another approach, involving geological field-evidence was taken by Wilson (1962), who considered the situation idealized in Fig.12.2. Lavas often intrude as dykes in older, sedimentary rocks thereby re-heating the contact-zones. If the re-heating is strong enough, the sedimentary rocks in the contact zones acquire TRM contemporaneous with the dyke itself. A reversed geomagnetic field will therefore be recorded as reversed dykes *and* reversed sedimentary rocks in the contact zone. A self-reversal either in the dyke or in the baked contact only will give opposite polarities. The polarities of 87 recorded baked

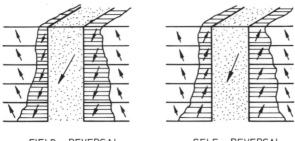

Fig.12.2. Coincidence of magnetic polarities of an intrusive rock and the baked (and remagnetized) intruded sediments indicates a field reversal. Opposite polarities indicate self-reversal. A statistical study of such baked contacts by Wilson (1962) provided compelling evidence for field reversals.

TABLE 12.I

Magnetic polarities of intrusive igneous rocks and baked country rock: data by Irving (1964) based on the original analysis by Wilson (1962)

Igneous polarity	Baked contact polarity	Number of cases
Normal	normal	34
Reversed	reversed	49
Oblique	oblique	2
Normal	reversed	2
Reversed	normal	0

contacts given in Table 12.I indicate only two self reversals out of the total presuming the probability of self-reversal in both intrusive and contact rocks to be small. A paradoxical case arises in the very extensive sequence of lavas of the Columbia Plateau, Oregon, in which Wilson and Watkins (1967) found a correlation between reversed polarity and oxidation; however, baked contact evidence from the same rocks indicates a field reversal rather than an oxidation-controlled self-reversal.

Probably more important to the theory of the origin of the geomagnetic field is the rock-magnetic evidence of the process of reversal. This has been seen both by intensive sampling of rapid sequences of lavas (Van Zijl et al., 1962) and in cores of deep-sea sediment (e.g., Ninkovitch et al., 1966). During the reversal process the field intensity is sharply diminished, as illustrated by the example in Fig.12.3, indicating that the dipole field vanishes and reappears with opposite polarity rather than rotating through 180°, while the weaker non-dipole features of the field probably persist with normal strength.

An important consequence of geomagnetic reversals arises from the process of seafloor spreading; the ocean ridges are linear sources of new ocean floor, from which upwelling basalt emerges to form oceanic crust. The crust spreads away from the ridges, eventually to be consumed at ocean trenches. (For reviews of this topic, see Isacks et al.,

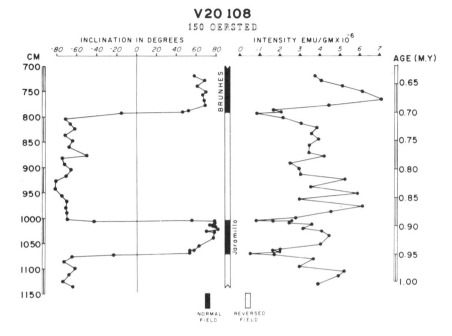

Fig.12.3. Detailed plot of the inclination and intensity of magnetization in specimens from part of an ocean sediment core, after partial demagnetization in 150 Oe. This demonstrates very well the diminution in field intensity during a reversal.
(Data from Ninkovitch et al., 1966, reproduced by permission, from the re-drawn figure of Opdyke, 1970.)

1968; Bullard, 1969.) As it is formed, the new ocean floor acquires remanence in the earth's field, but with a sequence of geomagnetic reversals the magnetic polarity of the ocean floor alternates, giving linear magnetic anomalies parallel to the ridges, as first realized by Vine and Matthews (1963). The anomaly pattern has been correlated with the sequence of geomagnetic reversals and used to estimate sea floor spreading rates (e.g., Heirtzler et al., 1968).

12.2 SELF-REVERSAL MECHANISMS

Detailed study of spontaneous self-reversal of magnetization in minerals began with a theoretical study by Néel (1951), who was prompted by J. W. Graham to apply his expertise in the magnetism of ferrites to the reversals problem. Néel considered four possible mechanisms, two requiring super-exchange interactions within magnetic minerals and the other two requiring only magnetostatic interactions between different minerals. Subsequent discussions (e.g., Uyeda, 1955; Stacey, 1963) have shown that the magnetostatic mechanisms are unpromising because the requirements on grain geometry and spontaneous magnetization, which are necessary for the interaction between grains (or lamellae within a grain) to be stronger than the external field, are difficult to meet. Although reversals resulting from magnetostatic interactions have occasionally been reported it is probable that no such cases exist. The classic example is the dacite from Mt. Haruna, Japan, which was shown in the laboratory to be self-reversing and to contain an intimate mixture of two magnetic minerals, an ilmeno-hematite and a titanomagnetite, indicating reversal by magnetostatic interaction (Nagata et al., 1952; Uyeda, 1955), but subsequent separation of the minerals revealed that the self-reversal was an intrinsic property of one of them alone and that the presence of titanomagnetite was irrelevant (Uyeda, 1958; Nagata and Uyeda, 1959). On the other hand, exchange interactions can be so strong that in one extreme case (Nagata and Uyeda, 1959) a synthetic ilmenite-hematite solid solution was found to acquire reversed TRM in all fields up to 16,000 Oe (Fig.12.4).

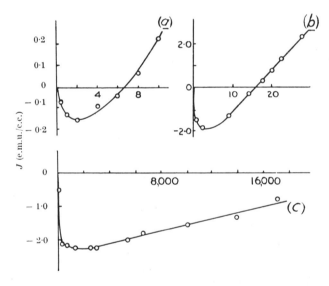

Fig.12.4. Reversed thermoremanence as a function of inducing field for (a). PTRM (350°C–250°C) for unseparated minerals of Haruna dacite; (b) total TRM of "constituent B", an ilmeno-hematite; and (c) total TRM of a laboratory-prepared ilmeno-hematite, $Fe_{1.52}Ti_{0.48}O_3$.
(Figure reproduced by permission, from Nagata and Uyeda, 1959.)

Several possible compositions of self-reversing minerals are discussed in Chapter 2. The best-documented self-reversing natural minerals are the hematite-ilmenite solid solutions (Fe_2O_3 –$FeTiO_3$). It was "component B", $Fe_{1.52}Ti_{0.48}O_3$, which was found to be responsible for self-reversal in the Haruna rock. Ishikawa and Syono (1963) confirmed a suggestion by Uyeda (1958) and favoured also by Carmichael (1961) that ionic ordering of the cations in ilmeno-hematite was responsible for self-reversals of TRM, but found that it was not simply a result of generating an ordered (low temperature) phase from a disordered (high temperature) one by ionic diffusion with reversed magnetization of the ordered phase relative to the disordered one. Reversals were only observed when a metastable, partially ordered phase was present; specimens either completely ordered or completely disordered did not have reversed TRM. They attributed the reversal to an antiparallel super-exchange interaction between the ordered phase and a metastable Ti-depleted zone of partial order. To be effective the ordered volumes must be small, so that the ratio of the area of surface interacting with the metastable phase to ordered volume is large. This requires fairly low temperature annealing ($\leqslant 600°C$); maintaining the temperature above $700°C$ causes the ordered regions to become too large (probably into the multidomain range) and the reversing property is lost. Westcott-Lewis and Parry (1971b) made the interesting observation that the self-reversing property of ilmeno-hematites disappeared for grains smaller than 4 μm, which acquire only normal thermoremanence. Their explanation is based on an argument that chemical inhomogeneities become important in small grains which do not contain sufficient of both the ordered and metastable phases. However, it is also possible that the properties are modified by internal stresses or grain surface effects. We note that in spite of this extremely well-documented case, self-reversal by exchange anisotropic coupling appears to be very rare, although positive exchange anisotropic coupling has been reported (Banerjee, 1966).

Some other possible self-reversal mechanisms are much more difficult to identify. Verhoogen (1956, 1962) considered the possibility of reversal by ionic reordering either following oxidation of titanomagnetites (1962) or due to order—disorder transitions involving impurities (1956) (in which the state of oxidation will also be important). The difficulty in recognizing such a situation is that it is an irreversible process and cannot be repeated by reheating and cooling in the laboratory. Fig.12.5 shows the range of com-

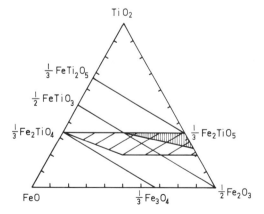

Fig.12.5. Ternary composition diagram of $FeO-Fe_2O_3-TiO_2$, showing the compositions of possible self-reversal, according to Verhoogen, 1962 (hatched area), and O'Reilly and Banerjee, 1967 (heavily shaded area).

positions of titanomaghemites which could be self-reversing according to Verhoogen. However O'Reilly and Banerjee (1967) argued that the appropriate composition range was much more restricted (to the heavily shaded area in Fig.12.5). Verhoogen had assumed that upon oxidation all Fe^{3+} ions would preferentially occupy tetrahedral sites, whereas O'Reilly and Banerjee (1965) had shown that the preference of Fe^{3+} for tetrahedral sites could be outweighed by the preference of Fe^{2+} for the same sites, due to the reduction in Coulomb energy. They also showed the initial distribution of ions assumed by Verhoogen to be in error. The range of self-reversing compositions allowed by the considerations of O'Reilly and Banerjee is so highly oxidized and therefore metastable, that it is unlikely to remain a single phase, as required for self-reversal by relative moment inversion of octahedral and tetrahedral lattices, but would exsolve into ilmenite and a low Ti titanomagnetite which would acquire normal magnetization.

The best known examples of Néel's N-type ferrimagnetics are all ternary ferrites with compositions quite unlike those recognized as significant in rock magnetism (e.g., $Li_{0.5}Fe^{3+}_{2.5-a}Cr_aO_4$ with $1 < a < 1.7$: Gorter and Schulkes, 1953). However, two examples of N-type ferrimagnetism appear in the rock magnetism literature. Schult (1968) found self-reversal of natural remanence in basalt specimens from Germany when cooled below 200°K; at the temperature of inversion of natural remanence the saturation magnetization showed a minimum, demonstrating the presence of an N-type mineral. But no specimens were found to give self-reversals above about 200°K. The only example of laboratory-reproducible N-type behaviour with an inversion temperature above 300°K was reported by Kropáček (1968) for a tin-substituted hematite embedded in a cassiterite rock.

Verhoogen (1956) also discussed the possibility of inversion of $(|A| - |B|)$ sublattice magnetizations by thermally reversible order—disorder transitions in tertiary or ternary ferrites. However, allowing that ionic diffusion rates may be appropriate for this process, no naturally occurring ferrites have been found with suitable compositions.

In view of the multiplicity of rather complex processes which may cause self-reversals, some not directly reproducible in the laboratory, a definite test for self-reversal of natural remanence is hardly possible. But several effects can suggest self-reversal and indicate rocks whose polarities require closer examination, such as physical or chemical differences between normal and reversely magnetized rocks in close proximity. The particular case of the correlation between reversed polarity and oxidation of the Columbia Plateau basalts, reported by Wilson and Watkins and mentioned in section *12.1*, necessitates caution in the interpretation of the polarities of these rocks. Other similar correlations have been reported, but so have absences of correlations and in spite of much detailed work, the explanation of this effect remains obscure. Domen (1969) has reported consistent differences between the demagnetization curves and Curie points of normally and reversely magnetized basalts from Kawajiri, Japan, and shown the existence of self-reversals in the reversed samples by comparison of thermal demagnetization curves (Fig.12.6). Minnibaev et al. (1966) reported measurements on ultrabasic volcanic rocks

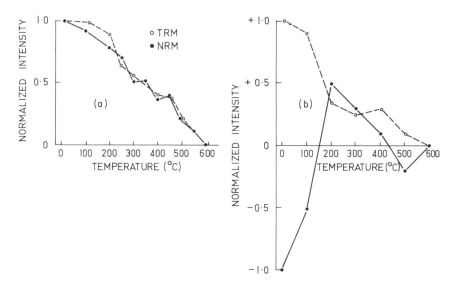

Fig.12.6. Comparison of thermal demagnetization curves for Kawajiri basalt samples with (a) normal and (b) reversed natural polarities. (After Domen, 1969.)

which gave normal TRM after a first heating but reversed TRM after the second and subsequent heatings. They compared alternating field demagnetizations of NRM, TRM and ARM and suggested that an indication of a self-reversing tendency was a greater resistance of ARM than TRM to demagnetization.

In spite of the multiplicity of possible mechanisms, demonstrated examples of self-reversal remain rare and the conditions required for self-reversal by those mechanisms which are understood are sufficiently stringent that the physical study of reversals can only be taken to confirm the paleomagnetic evidence that the great majority of reversed natural remanences indicate geomagnetic field reversals. Isolated reversed rocks must be treated with caution, but correlations of polarity between different rocks of the same age and field tests, such as that based on baked contacts (Table 12. I), give secure evidence of specific field reversals.

Chapter 13

MAGNETISM IN METEORITES

13.1 METEORITE TYPES AND METEORITIC IRON

Meteorites are of several types, all essentially of iron and stone in different proportions. They appear to be collision fragments of asteroidal bodies at least several tens and up to a few hundred kilometres in diameter. Since they evidently originated early in the development of the solar system they have been intensively studied in connection with its origin and by implication the origin of the earth, whose overall composition is believed to be similar to the meteoritic average. A monographic discussion of the whole subject is by Mason (1962) and the metal phases in stony-iron meteorites, which particularly concern us here, are considered in detail by Wood (1967).

Iron meteorites are composed almost entirely of nickel-iron with small amounts of troilite (FeS). They occur in large crystal sizes and are magnetically soft. Since this means that no particular geophysical significance can be attached to any magnetic remanence that they may have, detailed magnetic work on iron meteorites has not been reported. About 85% of meteorites are classed as *chondrites*, which are of stony composition, characterized by the occurrence of spherical grains or chondrules about 1 mm across, and normally containing also dispersed metal of composition similar to that found in iron meteorites. The iron-nickel grains are responsible for the magnetic properties, which are of particular interest because they are sufficiently coercive to retain remanence from their extra-terrestrial origin (section *13.2*). Ordinary chondrites contain no significant magnetite but there is a special class, the carbonaceous chondrites, characterized by the presence of carbon and carbon compounds, which are graded from Type I, with several percent of magnetite but no metal to Type III with little or no magnetite, but nickel-rich nickel-iron grains. They are evidently close in composition to the primitive material from which the terrestrial planets and meteorites accreted.

As in the case of iron meteorites the metal phases of the ordinary chondrites are not homogeneous nickel-iron alloy but exsolved nickel-rich and nickel-poor phases. The iron-rich end of the nickel-iron phase diagram is reproduced in Fig.13.1. At sufficiently high temperatures nickel-iron of all compositions has the face-centred cubic or γ structure, and in meteorites this form is known as *taenite*. Below the upper of the two phase boundaries in Fig.13.1, nucleation of centres of the body-centred cubic or α structure begins; meteoritic iron with this structure is known as *kamacite*. With slow cooling of a composition with about 12% Ni, which is typical of chondritic iron, the kamacite zones spread out and Ni withdraws by diffusion into the shrinking taenite zones, which become enriched in Ni to about 30%, leaving the kamacite with about 7% Ni. Eventually the diffusion becomes so sluggish that the phases become non-uniform. Diffusion across the

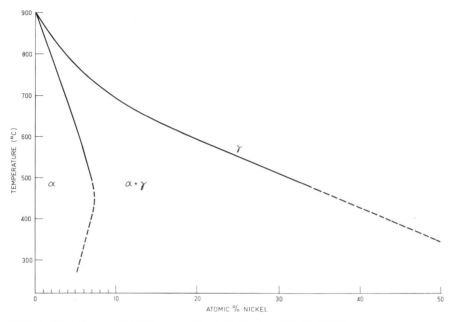

Fig.13.1. Phase diagram of nickel-iron according to Goldstein and Ogilvie (1965).

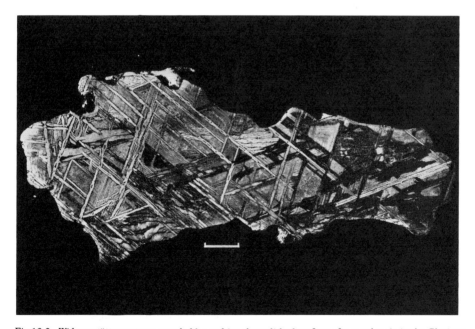

Fig.13.2. Widmanstätten pattern revealed by etching the polished surface of a metal grain in the Glorietta Mountain pallasite (stony-iron meteorite). The scale line is 1 cm.
(Photograph courtesy of J.F. Lovering.)

kamacite-taenite phase boundaries continues producing a slightly lower-than-kamacite average Ni content in the kamacite boundary and strongly enriched Ni in the taenite boundary. The exsolution pattern is rendered visible by etching a polished surface and is known as the Widmanstätten structure (Fig.13.2).

From a study of several ordinary chondrites Stacey et al. (1961) concluded that the magnetic properties were completely dominated by the α phase, kamacite, which is strongly magnetic. When heated it transforms to non-magnetic γ phase at a characteristic temperature which is a function of composition, generally about 750°C for chondritic kamacite, without going through a Curie point which would be slightly higher if the α phase could be stabilized to a sufficiently high temperature (as it is in pure iron). The disappearance of ferromagnetic properties is thus due to a phase change and not a normal Curie point. On re-cooling the γ phase reforms at about 650°C, 100°C lower than its disappearance. The natural remanence observed in chondrites must be explained in terms of this property of kamacite.

13.2 NATURAL REMANENCE IN CHONDRITIC METEORITES

Three independent experiments on different meteorites show that all of the twelve ordinary chondrites examined had stable natural magnetizations of magnitudes corresponding to thermoremanence induced in fields between 0.1 and 0.4 Oe[1] (Stacey et al., 1961; Weaving, 1962; Gus'kova, 1963). Comparison of natural remanences with laboratory-induced thermoremanences gave estimates of the field in which the natural remanences were induced, as summarized in Table 13.I. Strictly, neither the natural nor laboratory-induced moments are likely to have been entirely thermoremanent in the conventional sense, but involve a chemical remanent process, since the remanence appears to be induced at the $\gamma \to \alpha$ phase transition of kamacite during cooling of the chondrites. However, this is not important. What matters is that the natural remanences have the character of TRM or CRM associated with cooling from a temperature of 600+°C and can certainly not have been induced isothermally. This is particularly clear in Weaving's (1962) comparison of NRM with TRM and IRM for a sample from the centre of the Brewster meteorite (Fig.13.3). Both Weaving (1962) and Gus'kova (1963) used alternating field demagnetization to establish the character of the natural remanences; Stacey et al. (1961) used thermal demagnetization.

It is important to note that the central portion of a meteorite more than about 6 cm in diameter is not appreciably heated by its flight through the atmosphere (Lovering et al., 1966). Ablation of the surface keeps pace with the inward diffusion of heat, leaving only a 3 cm heated skin. Stacey et al. (1961) used only interior portions of observed falls

---

[1] In one case (Homestead, see Table 13.I) 0.9 Oe, but this is a result of calculation, not a thermoremanence experiment and could be in error by a factor of 2.

TABLE 13.I

Summary of published estimates of fields responsible for the primary remanences of chondrites. (The bracketed value is a theoretical estimate.)

Chondrite	Field intensity (Oe)	Reference
Ordinary chondrites		
Mt. Browne	0.25	
Farmington	0.18	Stacey et al. (1961)
Homestead	(0.9)	
Brewster	0.1	Weaving (1962)
Rakovka	0.4	
Mordvinovka	0.4	
Okhansk, no.2	0.3	
Okhansk, no.3	0.1	Gus'kova (1963)
Pultusk, no.1	0.2	
Pultusk, no.2	0.25	
Zhovtenevy Khutor, no.1350	0.15	
Zhovtenevy Khutor, no.199	0.2, 0.15	
Carbonaceous chondrites		
Mokoia	less than $5 \cdot 10^{-3}$	Stacey et al. (1961)
Allende	1.09	Butler (1972), Banerjee and Hargraves (1972)
Murchison	0.18	Banerjee and Hargraves (1972)
Orgeuil	0.67	

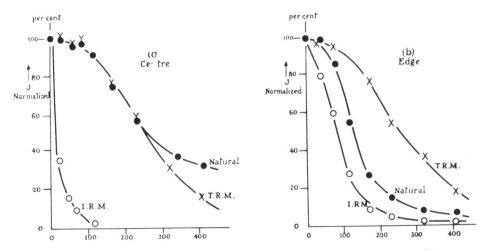

Fig.13.3. Comparison of natural remanence with laboratory-induced TRM and IRM for two samples of Brewster meteorite, one from the central portion and one from the edge.
(Figure by permission, from Weaving, 1962.)

which had no doubtful magnetic history. Thus we cannot avoid the conclusion that the natural remanences of chondrites are of extra-terrestrial origin.

The magnetic moments of carbonaceous chondrites are more variable. Stacey et al. (1961) found Mokoia to have a small moment which was so readily destroyed by heating that they concluded that it was terrestrially induced. However, recent work by Banerjee and Hargraves (1972) on three further carbonaceous chondrites (also listed in Table 13.I) shows that all three have stable moments of extra-terrestrial origin. Thus we have no basis for making a distinction between carbonaceous chondrites and ordinary chondrites in terms of their magnetic moments.

We also note that a special explanation is required for the very strong magnetization which Weaving (1962) found in the skin of Brewster meteorite. The fact that it was confined to the skin, but was much stronger than could be explained in terms of magnetization by the earth's field implies the presence of strong circulating currents within the heated skin (and perhaps due partly to the plasma immediately surrounding it during ablation).

The favoured explanation for the magnetization of the chondrites has changed in the last few years. Stacey (1967b) concluded that the necessary fields were specifically associated with the meteorite parent bodies, either by steady terrestrial-type dynamo action or as transient fields during break-up of the parent body, but later (Stacey, 1969): "The necessary field may have been associated with the solar nebula during the early stages of formation of the solar system." Objections to the steady parent-body field now appear overwhelming. Such a field requires a parent body with a large iron core still liquid, while the chondritic mantle was cool enough to have acquired remanence. We now know that the Widmanstätten structures of chondritic iron imply cooling so slow that the chondrites must have been buried some hundred kilometers deep (Wood, 1967), at which depth they must have been too hot to have been magnetic if the parent had a liquid core. Conversely if the core had solidified there would have been no planetary field.

The solar-field argument appeals to the evidence that a strong field was associated with the T-Tauri phase of the sun, when it was losing angular momentum by magnetic coupling to the surrounding disc of plasma (Alfvén, 1954; Sonnet et al., 1970). The assumption is therefore that the chondritic parent body or bodies had cooled below 600°C before the solar field was switched off. It is required that the field (or the relevant vector component of it) must have been consistently parallel to any rotational axis of the body in order that a steady magnetizing field should be apparent to the cooling chondritic material. Since the field is presumed to have been wound up into a spiral, the simplest way of satisfying the rotation requirement is for the body to have stopped rotating, i.e., to present a constant face to the sun (as the moon does to the earth). It is not easy to justify this.

Butler and Cox (1971) suggested that the natural remanences of chondrites were induced isothermally by a transient solar field and then hardened to more stable remanences by radiation damage caused by prolonged exposure to cosmic radiation. They found, as expected, a slight increase in coercivity of iron of meteoritic compositions when exposed to high neutron fluxes in a reactor. However, there are some serious difficulties also in the way of this mechanism of meteorite magnetization. Firstly, the remanence could not be increased by the irradiation, but merely stabilized. The proposal therefore requires a field of strength sufficient to induce the observed remanence isothermally and this is necessarily very much stronger than the field required to induce a comparable thermoremanence; the field which is believed to have existed at earth or asteroid distances from the sun during its T-Tauri phase is, however, compatible with the induction of thermoremanence. Secondly, irradiation of a solid produces point defects, whose effect on magnetic properties is slight compared with the dislocation multiplication caused by work-hardening. It therefore appears unlikely that irradiation contributes significantly to the observed stability of the remanences of chondrites.

Appendix 1

Table of the function $F'(a) = \int_0^1 x \tanh(ax)\,dx$

a	$F'(a)$	a	$F'(a)$	a	$F'(a)$	a	$F'(a)$	a	$F'(a)$
0.1	0.03327	1.1	0.3002	2.2	0.4218	4.5	0.4797	10.0	0.4959
0.2	0.06614	1.2	0.3176	2.4	0.4327	5.0	0.4836	11.0	0.4966
0.3	0.09824	1.3	0.3334	2.6	0.4417	5.5	0.4864	12.0	0.4971
0.4	0.12925	1.4	0.3476	2.8	0.4491	6.0	0.4886	13.0	0.4976
0.5	0.15888	1.5	0.3605	3.0	0.4553	6.5	0.4903	14.0	0.4979
0.6	0.18693	1.6	0.3722	3.2	0.4604	7.0	0.4916	15.0	0.4982
0.7	0.21324	1.7	0.3827	3.4	0.4648	7.5	0.4927	16.0	0.4984
0.8	0.23773	1.8	0.3922	3.6	0.4685	8.0	0.4936	17.0	0.4986
0.9	0.26038	1.9	0.4008	3.8	0.4717	8.5	0.4943	18.0	0.4987
1.0	0.28120	2.0	0.4085	4.0	0.4744	9.0	0.4949	19.0	0.4989
								20.0	0.4990

Appendix 2

Table of the function $F(a) = \int_0^1 \int_0^1 xy \tanh(axy)\,dx\,dy$

a	$F(a)$	a	$F(a)$	a	$F(a)$	a	$F(a)$	a	$F(a)$
0.1	0.01110	1.1	0.1079	2.2	0.1690	4.5	0.2165	10.0	0.2399
0.2	0.02212	1.2	0.1153	2.4	0.1761	5.0	0.2211	11.0	0.2413
0.3	0.03298	1.3	0.1224	2.6	0.1824	5.5	0.2248	12.0	0.2425
0.4	0.04362	1.4	0.1290	2.8	0.1880	6.0	0.2278	13.0	0.2434
0.5	0.05397	1.5	0.1352	3.0	0.1929	6.5	0.2303	14.0	0.2441
0.6	0.06398	1.6	0.1410	3.2	0.1973	7.0	0.2324	15.0	0.2448
0.7	0.07361	1.7	0.1465	3.4	0.2012	7.5	0.2342	16.0	0.2453
0.8	0.08283	1.8	0.1516	3.6	0.2046	8.0	0.2357	17.0	0.2458
0.9	0.09162	1.9	0.1564	3.8	0.2087	8.5	0.2370	18.0	0.2461
1.0	0.09997	2.0	0.1609	4.0	0.2106	9.0	0.2381	19.0	0.2465
								20.0	0.2468

Appendix 3

Table of the Langevin Function, $L(a) = \coth(a) - 1/a$

a	$L(a)$	a	$L(a)$	a	$L(a)$	a	$L(a)$	a	$L(a)$
0.1	0.03331	1.1	0.3401	2.2	0.5703	4.5	0.7780	10.0	0.9000
0.2	0.06649	1.2	0.3662	2.4	0.5999	5.0	0.8001	11.0	0.9091
0.3	0.09940	1.3	0.3912	2.6	0.6265	5.5	0.8182	12.0	0.9167
0.4	0.13193	1.4	0.4152	2.8	0.6503	6.0	0.8333	13.0	0.9231
0.5	0.16395	1.5	0.4381	3.0	0.6716	6.5	0.8462	14.0	0.9286
0.6	0.19536	1.6	0.4600	3.2	0.6908	7.0	0.8571	15.0	0.9333
0.7	0.22605	1.7	0.4808	3.4	0.7081	7.5	0.8667	16.0	0.9375
0.8	0.25594	1.8	0.5006	3.6	0.7237	8.0	0.8750	17.0	0.9412
0.9	0.28496	1.9	0.5195	3.8	0.7378	8.5	0.8824	18.0	0.9444
1.0	0.31304	2.0	0.5373	4.0	0.7507	9.0	0.8889	19.0	0.9474
								20.0	0.9500

Appendix 4

Table of the function $F''(a) = \dfrac{1}{a} \ln\left(\dfrac{\sinh a}{a} \right)$

a	$F''(a)$	a	$F''(a)$	a	$F''(a)$	a	$F''(a)$	a	$F''(a)$
0.1	0.01666	1.1	0.1765	2.2	0.3209	4.5	0.5117	10.0	0.7004
0.2	0.03329	1.2	0.1912	2.4	0.3430	5.0	0.5395	11.0	0.7190
0.3	0.04985	1.3	0.2056	2.6	0.3638	5.5	0.5640	12.0	0.7352
0.4	0.06632	1.4	0.2197	2.8	0.3834	6.0	0.5858	13.0	0.7494
0.5	0.08265	1.5	0.2335	3.0	0.4019	6.5	0.6054	14.0	0.7620
0.6	0.09883	1.6	0.2470	3.2	0.4194	7.0	0.6230	15.0	0.7733
0.7	0.11482	1.7	0.2602	3.4	0.4395	7.5	0.6389	16.0	0.7834
0.8	0.13060	1.8	0.2730	3.6	0.4514	8.0	0.6534	17.0	0.7926
0.9	0.14615	1.9	0.2855	3.8	0.4661	8.5	0.6667	18.0	0.8009
1.0	0.16144	2.0	0.2976	4.0	0.4801	9.0	0.6788	19.0	0.8085
								20.0	0.8156

BIBLIOGRAPHY

Ade-Hall, J.M., Wilson, R.L. and Smith, P.J., 1965. The petrology, Curie points and natural magnetizations of basic lavas. *Geophys. J.*, 9: 323.

Aharoni, A., Frei, E.H. and Schieber, M., 1962. Some properties of γ-Fe_2O_3 obtained by hydrogen reduction of α-Fe_2O_3. *J. Phys. Chem. Solids*, 23: 545.

Akimoto, S., 1962. Magnetic properties of $FeO-Fe_2O_3-TiO_2$ system as a basis of rock magnetism. *J. Phys. Soc. Japan*, 17 (Suppl. B 1): 706.

Akimoto, S., Katsura, T. and Yoshida, M., 1957. Magnetic properties of $TiFe_2O_4-Fe_3O_4$ system and their change with oxidation. *J. Geomagn. Geoelectr.*, 9: 165.

Alfvén, H., 1954. *The Origin of the Solar System*. Clarendon Press, Oxford.

Amar, H., 1958. Magnetization mechanism and domain structure of multidomain particles. *Phys. Rev.*, 111: 149.

Anderson, P.W., 1959. New approach to the theory of superexchange interactions. *Phys. Rev.*, 115: 2.

Anderson, P.W., 1963. Exchange in insulators: superexchange, direct exchange, and double exchange. In: G.T. Rado and H. Suhl (Editors), *Magnetism*, 1. Academic Press, New York, N.Y., p.25.

Anderson, P.W., Merritt, F.R., Remeika, J.P. and Yager, W.A., 1954. Magnetic resonance in α-Fe_2O_3. *Phys. Rev.*, 93: 717.

As, J.A. and Zijderveld, J.D.A., 1958. Magnetic cleaning of rocks in paleomagnetic research. *Geophys. J.*, 1: 308.

Bagina, O.L., 1966. Origin of the natural remanent magnetism of the redbed (Krasnotsvet) clays of the Gzhel' Stage in the Moscow basin. *Phys. Solid Earth*, 966: 809.

Balsley, J.R. and Buddington, A.F., 1960. Magnetic susceptibility anisotropy and fabric of some Adirondack granites and orthogneisses. *Am. J. Sci.*, 258 A: 6.

Banerjee, S.K., 1963a. An attempt to observe the basal plane anisotropy of hematite. *Phil. Mag.*, 8: 2119.

Banerjee, S.K., 1963b. *Magnetic Properties of Rocks and Minerals*. Thesis, Univ. of Cambridge, 156 pp.

Banerjee, S.K., 1966. Exchange anisotropy in intergrown maghemite and hematite. *Geophys. J.*, 10: 449.

Banerjee, S.K., 1969. Origin of weak ferromagnetism and remanence in natural cassiterite crystals. *J. Geophys. Res.*, 74: 3789.

Banerjee, S.K., 1971. Decay of marine magnetic anomalies by ferrous ion diffusion. *Nature*, 229: 181.

Banerjee, S.K. and Bartholin, H., 1970. Critical point behaviour of γ-Fe_2O_3. *I.E.E.E. Trans. Magn.*, MAG-6: 299.

Banerjee, S.K. and Hargraves, R.B., 1972. Natural remanent magnetizations of carbonaceous chondrites and the magnetic field in the early solar system *Earth Planet. Sci. Lett.*, 17: 110.

Banerjee, S.K., Johnson, C.E. and Krs, M., 1970. Mössbauer study to find the origin of weak ferromagnetism in cassiterite. *Nature*, 225: 173.

Banerjee, S.K., O'Reilly, W., Gibb, T.C. and Greenwood, N.N., 1967. The behavior of ferrous ions in iron-titanium spinels. *J. Phys. Chem. Solids*, 28: 1323.

Bean, C.P. and Livingston, J.D., 1959. Superparamagnetism. *J. Appl. Phys.*, 30 (Suppl.): 120 S.

Beetz, W., 1860. Ueber die inneren Vorgänge, welche die Magnetizierung bedingen. *Ann. Physik, Ser.2*, 140: 107.

Benard, J., 1939. Etude de la décomposition du protoxyde de fer et de solutions solide. *Ann. Chim.*, 12: 5.

Bertaut, F., 1952. La structure de la pyrrhotine Fe_7S_8. *Compt. Rend.*, 234: 1295.

Bhathal, R.S. and Stacey, F.D., 1969. Field-induced anisotropy of magnetic susceptibility in rocks. *Pure Appl. Geophys.*, 76: 123.

Bhathal, R.S., Gillingham, D.E.W. and Stacey, F.D., 1969. Thermal relaxation of magnetic domain misalignment in rocks. *Pure Appl. Geophys.*, 76: 130.

Bickford, L.R., Pappis, J. and Stull, J.L., 1955. Magnetostriction and permeability of magnetite and cobalt-substituted magnetite. *Phys. Rev.*, 99: 1210.

Blackett, P.M.S., 1962. On distinguishing self-reversed from field-reversed rocks. *J. Phys. Soc. Japan*, 17 (B-1): 699.
Blasse, G., 1964. Crystal chemistry and some magnetic properties of mixed metal oxides with spinel structure. *Philips Res. Rept. Suppl.*, 3: 139 pp.
Bozorth, R.M., 1951. *Ferromagnetism*. Van Nostrand, New York, N.Y., 968 pp.
Bozorth, R.M., Walsh, D.E. and Williams, A.J., 1957. Magnetization of ilmenite-hematite system at low temperatures. *Phys. Rev.*, 108: 157.
Brailsford, F., 1966. *Physical Principles of Magnetism*. Van Nostrand, London, 274 pp.
Breiner, S. and Kovach, R.L., 1967. Local geomagnetic events associated with displacements on the San Andreas fault. *Science*, 158: 116.
Brown, H.C., Khan, M.A. and Stacey, F.D., 1964. A search for flow structure in columnar basalt using magnetic anisotropy measurements. *Pure Appl. Geophys.*, 57: 61.
Brown, W.F., 1960. Single domain particles: new uses of old theorems. *Am. J. Phys.*, 28: 542.
Brown, W.F. and Johnson, C.E., 1962. Temperature variation of saturation magnetization of gamma-ferric oxide. *J. Appl. Phys.*, 33: 2752.
Buddington, A.F. and Lindsley, D.H., 1964. Iron-titanium oxide minerals and synthetic equivalents. *J. Petrol.*, 5: 318.
Bullard, E.C., 1969. The origin of the oceans. *Sci. Am.*, 221 (3): 66.
Butler, R.F., 1972. Natural remanent magnetization and thermomagnetic properties of the Allende meteorite. *Earth Planet Sci. Lett.*, 17: 120.
Butler, R.F. and Cox, A.V., 1971. A mechanism for producing magnetic remanence in meteorites and lunar samples by cosmic ray exposure. *Science*, 172: 939.
Callen, E.R. and Callen, H.B., 1963. Static magnetoelastic coupling in cubic crystals. *Phys. Rev.*, 129: 578.
Carmichael, C.M., 1959. Remanent magnetism of Allard Lake ilmenites. *Nature*, 183: 1239.
Carmichael, C.M., 1961. The magnetic properties of ilmenite-hematite crystals. *Proc. R. Soc.*, A 263: 508.
Carmichael, I.S.E. and Nicholls, J., 1967. Iron-titanium oxides and oxygen fugacities in volcanic rocks. *J. Geophys. Res.*, 72: 4665.
Chevallier, R., 1951. Propriétés magnétiques de l'oxyde ferrique rhomboédrique ($Fe_2O_3\alpha$). *J. Phys. Rad.*, 12: 172.
Chevallier, R. and Matthieu, S., 1943. Propriétés magnétiques des poudres d'hématites—influence des dimensions des grains. *Ann. Phys.*, 18: 258.
Cinader, G., Flanders, P.J. and Shtrikman, S., 1967. Magnetization and Mössbauer studies of the field dependence of the Morin transition in hematite. *Phys. Rev.*, 162: 419.
Cohen, J., Creer, K.M., Pauthenet, R. and Srivastava, K., 1962. Propriétés magnétiques des substances antiferromagnétiques en grains fins. *J. Phys. Soc. Japan*, 17 (Suppl. B-1): 685.
Collinson, D.W., 1965a. Origin of remanent magnetization in certain red sandstones. *Geophys. J.*, 9: 203.
Collinson, D.W., 1965b. Depositional remanent magnetization in sediments. *J. Geophys. Res.*, 70: 4663.
Collinson, D.W., Creer, K.M. and Runcorn, S.K., 1967. *Methods in Paleomagnetism (Developments in Solid Earth Geophysics, 3)*. Elsevier, Amsterdam, 609 pp.
Colombo, U., Gazzarrini, G., Lanzavecchia, G. and Sironi, G., 1964. Mechanisms in the first stage of oxidation of magnetites. *Nature*, 202: 175.
Cottrell, A.H., 1953. *Dislocations and Plastic Flow in Crystals*. Clarendon Press, Oxford, 223 pp.
Cox, A., 1969. Geomagnetic reversals. *Science*, 163: 237.
Cox, A. and Doell, R.R., 1960. Review of paleomagnetism. *Bull. Geol. Soc. Am.*, 71: 645.
Cox, A., Doell, R.R. and Dalrymple, G.B., 1963. Geomagnetic polarity epochs and pleistocene geochronometry. *Nature*, 198: 1049.
Creer, K.M., 1959. A. C. demagnetization of unstable Triassic Keuper marls from S.W. England. *Geophys. J.*, 2: 261.
Creer, K.M., 1961. Superparamagnetism in red sandstones. *Geophys. J.*, 5: 16.

Day, R., O'Reilly, W. and Banerjee, S.K., 1970. Rotational hysteresis study of oxidized basalts. *J. Geophys. Res.*, 75: 375.
De Boer, F. and Selwood, P.W., 1954. The activation energy for the solid state reaction $\gamma\text{-Fe}_2\text{O}_3 \to \alpha\text{-Fe}_2\text{O}_3$. *J. Am. Chem. Soc.*, 76: 3365.
Dickson, G.O., Everitt, C.W.F., Parry, L.G. and Stacey, F.D., 1966. Origin of thermoremanent magnetization. *Earth Planet. Sci. Lett.*, 1: 222.
Domen, H., 1969. An experimental study on the unstable natural remanent magnetization of rocks as a paleogeomagnetic fossil. *Bull. Fac. Educ. Yamaguchi Univ.*, 18 (2): 1.
Domenicali, C.A., 1950. Magnetic and electric properties of natural and synthetic single crystals of magnetite. *Phys. Rev.*, 78: 458.
Dunlop, D.J., 1968. Monodomain theory: experimental verification. *Science*, 162: 256.
Dunlop, D.J., 1970. Hematite: intrinsic and defect ferromagnetism. *Science*, 169: 858.
Dunlop, D.J., 1971. Magnetic properties of fine particle hematite. *Ann. Géophys.*, 27: 269.
Dunlop, D.J., 1972. Magnetite: behavior near the single-domain threshold. *Science*, 176: 41.
Dunlop, D.J., 1973a. Superparamagnetic and single domain threshold sizes in magnetite. *J. Geophys. Res.*, in press.
Dunlop, D.J., 1973b. Magnetic characteristics of submicroscopic magnetite and the origin of thermoremanent magnetization. In preparation.
Dunlop, D.J. and West, G.F., 1969. An experimental evaluation of single domain theories. *Rev. Geophys.*, 7: 709.
Dzyaloshinski, I., 1958. A thermodynamic theory of "weak" ferromagnetism of antiferromagnetics. *J. Phys. Chem. Solids*, 4: 241.
Evans, M.E. and McElhinny, M.W., 1969. An investigation of the origin of stable remanence in magnetite-bearing igneous rocks. *J. Geomagn. Geoelectr.*, 21: 757.
Everitt, C.W.F., 1962a. Thermoremanent magnetization. III. Theory of multidomain grains. *Phil. Mag.*, 7: 599.
Everitt, C.W.F., 1962b. Self-reversal of magnetization in a shale containing pyrrhotite. *Phil. Mag.*, 7: 831.
Ewing, J.A., 1914. *Magnetic Induction in Iron and Other Metals*. The Electrician Publishing Co., London, 393 pp.
Flanders, P.J. and Remeika, J.P., 1965. Magnetic properties of hematite single crystals. *Phil. Mag.*, 11: 1271.
Flanders, P.J. and Schuele, W.J., 1964. Temperature-dependent magnetic properties of hematite. *Proc. Int. Conf. Magn. Nottingham, 1964.* Inst. Physics and Phys. Soc. Lond., p. 594.
Fletcher, E.J. and Banerjee, S.K., 1969. High temperature dependence of single crystal anisotropy constants of titanomagnetites (abstr.). *EOS, Trans. Am. Geophys. Union*, 50: 132.
Friedel, J., 1964. *Dislocations*. Pergamon, New York, N.Y., 491 pp.
Fuller, M.D., 1963. Magnetic anisotropy and paleomagnetism. *J. Geophys. Res.*, 68: 293.
Fuller, M.D., 1970. Geophysical aspects of paleomagnetism. *Crit. Rev. Solid State Phys.*, 1: 137.
Gazzarrini, F. and Lanzavecchia, G., 1969. Role of crystal structure, defects and cationic diffusion on the oxidation and reduction processes of iron oxides at low temperatures. In: J. W. Mitchell et al. (Editors), *Reactivity of Solids.* Wiley, New York, N.Y., p. 57.
Gillingham, D.E.W., 1971. *Alternating Field Properties of Magnetite-bearing Rocks.* Thesis, Univ. Queensland, 259 pp.
Gillingham, D.E.W. and Stacey, F.D., 1971. Anhysteretic remanent magnetization (A.R.M.) in magnetite grains. *Pure Appl. Geophys.*, 91: 160.
Goldstein, J.A. and Ogilvie, R.E., 1965. A re-evaluation of the iron-rich portion of the Fe-Ni system. *Trans. Metal. Soc. AIME*, 233: 2083.
Goodenough, J.B., 1963. *Magnetism and the Chemical Bond*. Wiley-Interscience, New York, N.Y., 393 pp.
Gorter, E.W. and Schulkes, J.A., 1953. Reversal of spontaneous magnetization as a function of temperature in LiFeCr spinels. *Phys. Rev.*, 90: 487.
Graham, J.W., 1954. Magnetic susceptibility anisotropy, an unexploited petrofabric element. *Bull. Geol. Soc. Am.*, 65: 1257.

Graham, J.W., 1956. Paleomagnetism and magnetostriction. *J. Geophys. Res.*, 61: 735.
Green, R. and Irving, E., 1958. The paleomagnetism of the Cainozoic basalts from Australia. *Proc. R. Soc. Vic.*, 70: 1.
Griffiths, D.H., King, R.F., Rees, A.I. and Wright, A.E., 1960. The remanent magnetism of some recent varved sediments. *Proc. R. Soc.*, A 256: 359.
Grommé, C.S., Wright, T.L. and Peck, D.L., 1969. Magnetic properties and oxidation of iron-titanium oxide minerals in Alae and Makaopuhi lava lakes, Hawaii. *J. Geophys. Res.*, 74: 5277.
Gus'kova, Ye.G., 1963. Investigation of natural remanent magnetization of stony meteorites. *Geomagn. Aeronomy*, 3: 308.
Haigh, G., 1957. Observations on the magnetic transition in hematite at $-15°C$. *Phil. Mag.*, 2: 877.
Haigh, G., 1958. The process of magnetization by chemical change. *Phil. Mag.*, 3: 267.
Hamilton, N. and King, R.F., 1964. Comparison of the bedding errors of artificially and naturally deposited sediments with those predicted from a simple model. *Geophys. J.*, 8: 370.
Hanuš, V. and Krs, M., 1965. Some minerals as new natural models of ferrites. *Pure Appl. Geophys.*, 62: 129.
Hargraves, R.B. and Young, W.M., 1969. Source of stable remanent magnetism in Lambertville diabase. *Am. J. Sci.*, 267: 1161.
Harrison, C.G.A., 1966. The paleomagnetism of deep sea sediments. *J. Geophys. Res.*, 71: 3033.
Hedley, I.G., 1968. Chemical remanent magnetization of the FeOOH, Fe_2O_3 system. *Phys. Earth Planet. Interiors*, 1: 103.
Heirtzler, J.R., Dickson, G.O., Herron, E.M., Pitman, W.C. and Le Pichon, X., 1968. Marine magnetic anomalies, geomagnetic field reversals and motions of the ocean floor and continents. *J. Geophys. Res.*, 73: 2119.
Heitler, W., 1945. *Elementary Wave Mechanics*. Clarendon Press, Oxford, 136 pp.
Holz, A., 1970. Formation of reversed domains in plate-shaped ferrite particles. *J. Appl. Phys.*, 41: 1095.
Howell, L.G., Martinez, J.D., Frosch, A. and Statham, E.H., 1960. A note on chemical magnetization of rocks. *Geophysics*, 25: 1094.
Irving, E., 1964. *Paleomagnetism and its Application to Geological and Geophysical Problems*. Wiley-Interscience, New York, N.Y., 399 pp.
Irving, E. and Major, A., 1964. Post-depositional detrital remanent magnetization in a synthetic sediment. *Sedimentology*, 3 : 135.
Isacks, B., Oliver, J. and Sykes, L.R., 1968. Seismology and the new global tectonics. *J. Geophys. Res.*, 73: 5855.
Ishikawa, Y., 1967. Magnetic properties of a single crystal of Fe_2TiO_4. *Phys. Lett.*, 24A: 725.
Ishikawa, Y. and Akimoto, S., 1958. Magnetic property and crystal chemistry of ilmenite ($MeTiO_3$) and hematite (αFe_2O_3) system, 2. Magnetic property. *J. Phys. Soc. Japan*, 13: 1298.
Ishikawa, Y. and Syono, Y., 1963. Order–disorder transformation and reverse thermo-remanent magnetism in the $FeTiO_3-Fe_2O_3$ system. *J. Phys. Chem. Solids*, 24: 517.
Jacobs, I.S. and Bean, C.P., 1955. An approach to elongated fine-particle magnets. *Phys. Rev.*, 100: 1060.
Johnson, E.A., Murphy, T. and Torreson, O.W., 1948. Pre-history of the earth's magnetic field. *Terr. Magn. Atm. Electr.*, 53: 349.
Johnston, M.J.S., 1970. *The Volcano-Magnetic Effect*. Thesis, Univ. Queensland, 210 pp.
Johnston, M.J.S. and Stacey, F.D., 1969a. Volcanomagnetic effect observed on Mt. Ruapehu, New Zealand. *J. Geophys. Res.*, 74: 6541.
Johnston, M.J.S. and Stacey, F.D., 1969b. Transient magnetic anomalies accompanying volcanic eruptions in New Zealand. *Nature*, 224: 1289.
Kapitsa, S.P., 1955. Magnetic properties of eruptive rocks exposed to mechanical stresses. *Izv. Akad. Nauk S.S.S.R., Ser. Geofiz.*, 6: 489.
Kawai, N., Kume, S. and Sasajima, S., 1954. Magnetism of rocks and solid phase transformation in ferromagnetic minerals. *Proc. Japan Acad.*, 30: 588, 864.
Kawai, N., Ono, F. and Hirooka, K., 1968. A new explanation for the magnetic memory of α-Fe_2O_3 on the basis of a negative pressure effect on the Morin transition point. *J. Appl. Phys.*, 39: 712.

Kern, J.W., 1961a. Effects of moderate stresses on directions of thermoremanent magnetization. *J. Geophys. Res.*, 66: 3801.
Kern, J.W., 1961b. Effect of stress on the susceptibility and magnetization of a partially magnetized multidomain system. *J. Geophys. Res.*, 66: 3807.
Khramov, A.N., 1968. Orientational magnetization of finely dispersed sediments. *Phys. Solid Earth*, 1968(1): 63.
King, R.F., 1955. The remanent magnetism of artificially deposited sediments. *Mon. Notices R. Astron. Soc. Geophys. Suppl.*, 7: 115.
King, R.F., 1966. The magnetic fabric of some Irish granites. *Geol. J.*, 5: 43.
King, R.F. and Rees, A.I., 1962. The measurement of the anisotropy of magnetic susceptibility of rocks by the torque method. *J. Geophys. Res.* 67: 1565.
King, R.F. and Rees, A.I., 1966. Detrital magnetism in sediments: an examination of some theoretical models. *J. Geophys. Res.*, 71: 561.
Kittel, C., 1949. Physical theory of ferromagnetic domains. *Rev. Mod. Phys.*, 21: 541.
Kittel, C., 1971. *Introduction to Solid-state Physics*. Wiley, New York, N.Y., 4th ed., 766 pp.
Kobayashi, K., 1959. Chemical remanent magnetization of ferromagnetic minerals and its application to rock magnetism. *J. Geomagn. Geoelectr.*, 10: 99.
Kobayashi, K., 1961. An experimental demonstration of the production of chemical remanent magnetization with Cu-Co alloy. *J. Geomagn. Geolectr.*, 12: 148.
Kobayashi, K., 1962. Magnetization-blocking process by volume development of ferromagnetic fine particles. *J. Phys. Soc. Japan*, 17 (Suppl. B-I): 695.
Kobayashi, K. and Fuller, M.D., 1968. Stable remanence and memory of multidomain materials with special reference to magnetite. *Phil. Mag.*, 18: 601.
Koenigsberger, J.G., 1938. Natural residual magnetism of eruptive rocks. *Terr. Magn. Atm. Electr.*, 43: 119, 299.
Kramers, H.A., 1934. The interaction between the magnetogenic atoms in a paramagnetic crystal. *Physica*, 1: 182 (in French).
Kropáček, V., 1968. Self-reversal of spontaneous magnetization of natural cassiterite. *Stud. Geophys. Geodet. Českoslov. Akad. Ved.*, 12: 108.
Kumagai, H., Abe, H., Ôno, K., Hayashi, I., Shimada, J. and Iwanaga, K., 1955. Frequency dependence of magnetic resonance in α-Fe_2O_3. *Phys. Rev.*, 99: 1116.
Kündig, W., Bömmel, H., Constabaris, G. and Lindquist, R.H., 1966. Some properties of supported small α-Fe_2O_3 particles determined with the Mössbauer effect. *Phys. Rev.*, 142: 327.
Kushiro, I., 1960. $\gamma \rightarrow \alpha$ transition in Fe_2O_3 with pressure. *J. Geomagn. Geoelectr.*, 11: 148.
Lewis, R.R. and Senftle, F.E., 1966. The source of ferromagnetism in zircon. *Am. Mineralogist*, 51: 1467.
Lovering, J.F., Parry, L.G. and Jaeger, J.C., 1960. Temperatures and mass losses in iron meteorites during ablation in the earth's atmosphere. *Geochim. Cosmochim. Acta*, 19: 156.
Lowrie, W. and Fuller, M., 1971. On the alternating-field demagnetization characteristics of multidomain thermoremanence in magnetite. *J. Geophys. Res.*, 76: 6339.
Mason, B., 1962. *Meteorites*. Wiley, New York, N.Y., 274 pp.
McDougall, I. and Tarling, D.H., 1963. Dating of polarity zones in the Hawaiian islands. *Nature*, 200: 54.
Merrill, R., 1968. A possible source for the coercivity of ilmenite-hematite minerals. *J. Geomagn. Geoelectr.*, 20: 181.
Merrill, R.T., 1970. Low-temperature treatments of magnetite and magnetite-bearing rocks. *J. Geophys. Res.*, 75: 3343.
Michel, A. and Chaudron, G., 1935. Etude du sesquioxyde de fer cubique stabilisé. *Compt. Rend.*, 201: 1191.
Minnibaev, R.A., Myasnikov, V.S. and Petrova, G.N., 1966. On a case of self-reversal of the remanent magnetization. *Phys. Solid Earth*, 1966: 536.
Mizushima, K. and Iida, S., 1966. Effective in-plane anisotropy field in α-Fe_2O_3. *J. Phys. Soc. Japan*, 21: 1521.

Moriya, T., 1960. Anisotropic exchange interaction and weak ferromagnetism. *Phys. Rev.*, 120: 91.
Morrish, A.H., 1965. *The Physical Principles of Magnetism*. Wiley, New York, N.Y., 680 pp.
Morrish, A.H. and Yu, S.P., 1955. Dependence of coercive force on the density of some iron oxide powders. *J. Appl. Phys.*, 26: 1049.
Nagata, T., 1953, 1961. *Rock Magnetism*. Maruzen, Tokyo, 1st ed. 1953, 225 pp.
Nagata, T., 1970. Basic magnetic properties of rocks under the effects of mechanical stresses. *Tectonophysics*, 9: 167.
Nagata, T. and Akimoto, S., 1956. Magnetic properties of ferromagnetic ilmenites. *Geofis. Pura Appl.*, 34: 36.
Nagata, T. and Kinoshita, H., 1967. Effect of hydrostatic pressure on magnetostriction and magnetocrystalline anisotropy of magnetite. *Phys. Earth Planet. Interiors*, 1: 44.
Nagata, T. and Uyeda, S., 1959. Exchange interaction as a cause of reverse thermoremanent magnetism. *Nature*, 184: 890.
Nagata, T., Uyeda, S. and Akimoto, S., 1952. Self-reversal of thermoremanent magnetism of igneous rocks. *J. Geomagn. Geoelectr.*, 4: 22.
Nathans, R., Pickart, S.J., Alperin, H.J. and Brown, P.J., 1964. *Phys. Rev.*, 136A: 1641.
Néel, L., 1940. Champ moléculaire, aimantation à saturation et constantes de Curie des éléments de transition et de leurs alliages. In: Institut International de Coopération Intellectuelle. *Le Magnétisme*, 2. *Ferromagnétisme*. Collection Scientifique, Paris, pp. 65–164.
Néel, L., 1944. Effet des cavités et des inclusions sur le champ coercitif. *Cahiers Physique*, 25: 21
Néel, L., 1948. Magnetic properties of ferrites: ferrimagnetism and antiferromagnetism. *Ann. Phys.*, 3: 137.
Néel, L., 1951. L'inversion de l'aimantation permanente des roches. *Ann. Geophys.*, 7: 90.
Néel, L., 1955. Some theoretical aspects of rock magnetism. *Adv. Phys.*, 4: 191.
Néel, L., 1962. Propriétés magnétiques des grains fins antiferromagnétiques: superparamagnétisme et superantiferromagnétisme. *J. Phys. Soc. Japan*, 17 (Suppl. B-I): 676.
Néel, L. and Pauthenet, R., 1952. Etude thermomagnétique d'un monocristal de $Fe_2O_3\alpha$. *Compt. Rend.*, 234: 2172.
Ninkovitch, D., Opdyke, N., Heezen, B.C. and Foster, J.H., 1966. Paleomagnetic stratigraphy, rates of deposition and tephrachronology in North Pacific deep-sea sediments. *Earth Planet. Sci. Lett.*, 1: 476.
Ohnaka, M. and Kinoshita, H., 1968a. Effect of uniaxial compression on remanent magnetization. *J. Geomagn. Geoelectr.*, 20: 93.
Ohnaka, M. and Kinoshita, H., 1968b. Effect of axial stress upon initial susceptibility of an assemblage of fine grains of $Fe_2TiO_4-Fe_3O_4$ solid solution series. *J. Geomagn. Geoelectr.*, 20: 107.
Opdyke, N.D., 1970. Paleomagnetism. In: A.E. Maxwell (Editor), '*The Sea*', 4 (Part 1). Wiley-Interscience, New York, p. 157.
O'Reilly, W. and Banerjee, S.K., 1965. Cation distribution in titanomagnetites $(1-x)$ Fe_3O_4-x Fe_2TiO_4. *Phys. Lett.*, 17: 237.
O'Reilly, W. and Banerjee, S.K., 1967. Mechanism of oxidation in titanomagnetites, a magnetic study. *Mineral. Mag.*, 36: 29.
O'Reilly, W. and Readman, P.W., 1971. The preparation and unmixing of cation deficient titanomagnetites. *Z. Geophys.*, 37: 321.
Osborn, J.A., 1945. Demagnetizing factors of the general ellipsoid. *Phys. Rev.*, 67: 351.
Ozima, M. and Larson E.E., 1970. Low- and high-temperature oxidation of titanomagnetite in relation to irreversible changes in the magnetic properties of submarine basalts. *J. Geophys. Res.*, 75: 1003.
Ozima, M. and Sakamoto, N., 1971. Magnetic properties of synthesized titanomaghemite. *J. Geophys. Res.*, 76: 7035.
Ozima, Minoru and Ozima, Mituko, 1965. Origin of thermoremanent magnetization. *J. Geophys. Res.*, 70: 1363.
Ozima, Minoru and Ozima, Mituko, 1972. Activation energy of unmixing of titanomaghemite. *Phys. Earth Planet. Interiors*, 5: 87.

Ozima, Minoru, Ozima, Mituko and Nagata, T., 1964. Low temperature treatment as an effective means of "magnetic cleaning" of natural remanent magnetization. *J. Geomagn. Geoelectr.*, 16: 37.
Parry, L.G., 1965. Magnetic properties of dispersed magnetite powders. *Phil. Mag.*, 11: 303.
Patton, B.J. and Fitch, J.L., 1962. Anhysteretic magnetization in small steady fields. *J. Geophys. Res.*, 67: 307.
Pauthenet, R., 1950. Variation thermique de l'aimantation spontanée des ferrites de nickel, cobalt, fer et manganèse. *Compt. Rend.*, 230: 1842.
Porath, H., 1968. Stress-induced magnetic anisotropy in natural single crystals of hematite. *Phil. Mag.*, 17: 603.
Porath, H. and Raleigh, C.B., 1967. An origin of the triaxial basal-plane anisotropy in hematite crystals. *J. Appl. Phys.*, 38: 2401.
Porath, H., Stacey, F.D. and Cheam, A.S., 1966. The choice of specimen shape for magnetic anisotropy measurements on rocks. *Earth Planet. Sci. Lett.*, 1: 92.
Pucher, R., 1969. Relative stability of chemical and thermal remanence in synthetic ferrites. *Earth Planet. Sci. Lett.*, 6: 107.
Readman, P.W., O'Reilly, W. and Banerjee, S.K., 1967. An explanation of the magnetic properties of Fe_2TiO_4. *Phys. Lett.*, 25A: 446.
Rees, A.I., 1961. The effect of water currents on the magnetic remanence and anisotropy of susceptibility of some sediments. *Geophys. J.*, 5: 235.
Riste, T. and Tenzer, L., 1961. A neutron diffraction study of the temperature variation of the spontaneous sublattice magnetization of ferrites and the Néel theory of ferrimagnetism. *J. Phys. Chem. Solids*, 19: 117.
Roquet, J., 1954. Sur les rémanences magnétiques des oxydes de fer et leur intérêt en géomagnétisme. *Ann. Geophys.*, 10: 282.
Rusakov, O.M., 1966. The stability of the oriented remanent magnetization. *Phys. Solid Earth*, 1966 (6): 408.
Sakamoto, N., Ince, P. and O'Reilly, W., 1968. Effect of wet grinding on the oxidation of titanomagnetites. *Geophys. J.*, 15: 509.
Schroeer, D., 1968. The quadrupole interaction in α-Fe_2O_3 microcrystals. *Phys. Lett.*, 27A: 507.
Schroeer, D. and Nininger, R.C., 1967. Morin transition in α-Fe_2O_3 microcrystals. *Phys. Rev. Lett.*, 19: 632.
Schult, A., 1968. Self-reversal of magnetization and chemical composition of titanomagnetites in basalts. *Earth Planet. Sci. Lett.*, 4: 1968.
Schwartz, E.J., 1968. Magnetic phases in natural pyrrhotite, $Fe_{0.89}S$ and $Fe_{0.91}S$. *J. Geomagn. Geoelectr.*, 20: 67.
Searle, C.W. and Morrish, A.H., 1966. A three sub-lattice theory of weakly ferromagnetic $\alpha M_\delta^{z^+} Fe_\delta^{z^+} Fe_{2(1-\delta)}^{3+} O_3$. *J. Appl. Phys.*, 37: 1141.
Shamsi, S. and Stacey, F.D., 1969. Dislocation models and seismomagnetic calculations for California 1906 and Alaska 1964 earthquakes. *Bull. Seismol. Soc. Am.*, 59: 1435.
Shimizu, Y., 1960. Magnetic viscosity of magnetite. *J. Geomagn. Geoelectr.*, 11: 125.
Shive, P.N., 1969. The effect of internal stress on the thermoremanence of nickel. *J. Geophys. Res.*, 74: 3781.
Shive, P.N. and Butler, R.F., 1969. Stresses and magnetostrictive effects of lamellae in the titanomagnetite and ilmenohematite series. *J. Geomagn. Geoelectr.*, 21: 781.
Shull, C.G., Strauser, W.A. and Wollan, E.O., 1951. Neutron diffraction by paramagnetic and antiferromagnetic substances. *Phys. Rev.*, 83: 333.
Smart, J., 1955. The Néel theory of ferrimagnetism. *Am. J. Phys.*, 23: 356.
Smit, J. and Wijn, H.P.J., 1959. *Ferrites*. Wiley-Interscience, New York, N.Y., 369 pp.
Smith, P.J., 1971. Field reversal or self-reversal? *Nature*, 229: 378.
Soffel, H.C., 1969. The origin of thermoremanent magnetization of two basalts containing homogeneous single-phase titanomagnetite. *Earth Planet. Sci. Lett.*, 7: 201.
Sonnet, C.P., Colburn, D.S., Schwarz, K. and Keil, K., 1970. The melting of asteroidal-sized bodies by unipolar dynamo induction from a primordial T Tauri sun. *Astrophys. Space Sci.*, 7: 446.

Stacey, F.D., 1958. Thermoremanent magnetization (TRM) of multidomain grains in igneous rocks. *Phil. Mag.*, 3: 1391.
Stacey, F.D., 1959. A domain theory of magnetic grains in rocks. *Phil. Mag.*, 4: 594.
Stacey, F.D., 1960a. Stress-induced magnetic anisotropy of rocks. *Nature*, 188: 134.
Stacey, F.D., 1960b. Magnetic anisotropy of igneous rocks. *J. Geophys. Res.*, 65: 2420.
Stacey, F.D., 1961. Theory of the magnetic properties of igneous rocks in alternating fields. *Phil. Mag.*, 6: 1241.
Stacey, F.D., 1962a. A generalized theory of thermoremanence, covering the transition from single-domain to multidomain magnetic grains. *Phil. Mag.*, 7: 1887.
Stacey, F.D., 1962b. Theory of the magnetic susceptibility of stressed rock. *Phil. Mag.*, 7: 551.
Stacey, F.D., 1963. The physical theory of rock magnetism. *Adv. Phys.*, 12: 45.
Stacey, F.D., 1964. The seismomagnetic effect. *Pure Appl. Geophys.*, 58: 5.
Stacey, F.D., 1967a. The Koenigsberger ratio and the nature of thermoremanence in igneous rocks. *Earth Planet. Sci. Lett.*, 2: 67.
Stacey, F.D., 1967b. Paleomagnetism of meteorites. In: S.K. Runcorn (Editor), *International Dictionary of Geophysics*. Pergamon, Oxford, p. 1141.
Stacey, F.D., 1969. *Physics of the Earth*. Wiley, New York, N.Y., 324 pp.
Stacey, F.D., 1972. On the role of Brownian motion in the control of detrital remanent magnetization in sediments. *Pure Appl. Geophys.*, 98: 139.
Stacey, F.D. and Johnston, M.J.S., 1972. Theory of the piezomagnetic effect in titanomagnetite-bearing rocks. *Pure Appl. Geophys.*, 97: 146.
Stacey, F.D. and Wise, K.N., 1967. Crystal dislocations and coercivity in fine-grained magnetite. *Aust. J. Phys.*, 20: 507.
Stacey, F.D., Joplin, G. and Lindsay, J., 1960. Magnetic anisotropy and fabric of some foliated rocks from S.E. Australia. *Pure Appl. Geophys.*, 47: 30.
Stacey, F.D., Lovering, J.E. and Parry, L.G., 1961. Thermomagnetic properties, natural magnetic moments and magnetic anisotropies of some chondritic meteorites. *J. Geophys. Res.*, 66: 1523.
Stacey, F.D., Barr, K.G. and Robson, G.R., 1965. The volcanomagnetic effect. *Pure Appl. Geophys.*, 62: 96.
Stephenson, A., 1969. The temperature-dependent cation distribution in titanomagnetites. *Geophys. J.*, 18: 199.
Stoner, E.C., 1945. Demagnetizing factors for ellipsoids. *Phil. Mag.*, 7 (36): 803.
Stoner, E.C. and Wohlfarth, E.P., 1948. A mechanism of magnetic hysteresis in heterogeneous alloys. *Phil. Trans. R. Soc.*, A 240: 599.
Stott, P.M. and Stacey, F.D., 1959. Magnetostriction and paleomagnetism of igneous rocks. *Nature*, 183: 384.
Stott, P.M. and Stacey, F.D., 1960. Magnetostriction and paleomagnetism of igneous rocks. *J. Geophys. Res.*, 65: 2419.
Strangway, D.W., 1961. Magnetic properties of diabase dikes. *J. Geophys. Res.*, 66: 3021.
Strangway, D.W., Larson, E.E. and Goldstein, M., 1968. A possible cause of high magnetic stability in volcanic rocks. *J. Geophys. Res.*, 73: 3787.
Sunagawa, I. and Flanders, P.J., 1965. Structural and magnetic studies in hematite single crystals. *Phil. Mag.*, 11: 747.
Syono, Y., 1965. Magnetocrystalline anisotropy and magnetostriction of $Fe_3O_4 - Fe_2TiO_4$ series, with special application to rock magnetism. *Japan. J. Geophys.*, 4: 71.
Tasaki, A. and Iida, S., 1963. Magnetic properties of a synthetic single crystal of $\alpha\text{-}Fe_2O_3$. *J. Phys. Soc. Japan*, 18: 1148.
Thellier, E. and Thellier, O., 1941. Sur les variations thermiques de l'aimantation thermoremanente des terres cuites. *Compt. Rend.*, 213: 59.
Torreson, O.W., Murphy, T. and Graham, J.W., 1949. Magnetic polarization of sedimentary rocks and the earth's magnetic history. *J. Geophys. Res.*, 54: 111.
Tropin, Yu. D. and Vlasov, A.Ya., 1966. Some aspects of the theory of magnetic viscosity of rocks. *Phys. Solid Earth*, 1966 (5): 325.

Uyeda, S., 1955. Magnetic interactions between ferromagnetic minerals contained in rocks. *J. Geomagn. Geoelectr.*, 7: 9.

Uyeda, S., 1958. Thermoremanent magnetism as a medium of palaeomagnetism, with special reference to reverse thermoremanent magnetism. *Japan. J. Geophys.*, 2: 1.

Uyeda, S., Fuller, M.D., Belshe, J.C. and Girdler, R.W., 1963. Anisotropy of magnetic susceptibility of rocks and minerals. *J. Geophys. Res.*, 68: 279.

Van der Woude, F., Sawatzky, G.A. and Morrish, A.H., 1968. Relation between hyperfine magnetic fields and sublattice magnetizations in Fe_3O_4. *Phys. Rev.*, 167: 533.

Van Vleck, J.H., 1932. *The Theory of Electric and Magnetic Susceptibilities*. Clarendon Press, Oxford, 384 pp.

Van Vleck, J.H., 1937. On the anisotropy of cubic ferromagnetic crystals. *Phys. Rev.*, 52: 1178.

Van Vleck, J.H., 1945. A survey of the theory of ferromagnetism. *Rev. Mod. Phys.*, 17: 27.

Van Zijl, J.S.V., Graham, K.W.T. and Hales, A.L., 1962. The palaeomagnetism of the Stromberg lavas of South Africa. *Geophys. J.*, 7: 23, 169.

Verhoogen, J., 1956. Ionic ordering and self-reversal in impure magnetites. *J. Geophys. Res.*, 61: 201.

Verhoogen, J., 1959. The origin of thermoremanent magnetization. *J. Geophys. Res.*, 64: 2441.

Verhoogen, J., 1962. Oxidation of iron-titanium oxides in igneous rocks. *J. Geol.*, 70: 168.

Verwey, E.J.W. and Haayman, P.W., 1941. Electronic conductivity and transition point in magnetite. *Physica*, 8: 979.

Verwey, E.J., Haayman, P.W. and Romeijn, F.C., 1947. Physical properties and cation arrangements of oxides with spinel structure. *J. Chem. Phys.*, 15: 181.

Vine, F.J. and Matthews, D.H., 1963. Magnetic anomalies over ocean ridges. *Nature*, 199: 947.

Vlasov, A. Ya, Kovalenko, G.V. and Chikhachev, V.A., 1967a. Temperature-induced phase and magnetic transitions of iron hydroxide. *Phys. Solid Earth*, 1967: 661.

Vlasov, A. Ya, Kovalenko, G.V. and Fedoseeva, N.V., 1967b. The rotational magnetic hysteresis of single hematite crystals in artificial sediments containing magnetite and hematite particles. *Phys. Solid Earth*, 1967: 129.

Wasilewski, P.J., 1969. Thermochemical remanent magnetization (TCRM) in basaltic rocks: experimental characteristics. *J. Geomagn. Geoelectr.*, 21: 595.

Weaving, B., 1962. The magnetic properties of the Brewster meteorite. *Geophys. J.*, 7: 203.

Weiss, P., 1907. L'hypothèse du champ moléculaire et la propriété ferromagnétique. *J. Physique*, 6: 661.

Westcott-Lewis, M.F., 1971. Grain-size dependence of thermoremanence in ilmenite-haematites. *Earth Planet. Sci. Lett.*, 12: 124.

Westcott-Lewis, M.F. and Parry, L.G., 1971a. Magnetism in rhombohedral iron-titanium oxides. *Aust. J. Phys.*, 24: 719.

Westcott-Lewis, M.F. and Parry, L.G., 1971b. Thermoremanence in synthetic rhombohedral iron-titanium oxides. *Aust. J. Phys.*, 24: 735.

Whitworth, R.W. and Stopes-Roe, H.V., 1971. Experimental demonstration that the couple on a bar magnet depends on H, not B. *Nature*, 234: 31.

Wilson, E., 1922. On the susceptibility of feebly magnetic bodies as affected by compression. *Proc. R. Soc., London*, A101: 445.

Wilson, R.L., 1961. Palaeomagnetism in Northern Ireland. Part II. On the reality of a reversal in the earth's magnetic field. *Geophys. J.*, 5: 59.

Wilson, R.L., 1962. The palaeomagnetism of baked contract rocks and reversals of the earth's magnetic field. *Geophys. J.*, 7: 194.

Wilson, R.L. and Watkins, N.D., 1967. Correlation of petrology and natural magnetic polarity in Columbia Plateau basalts. *Geophys. J.*, 12: 405.

Wolf, W.P., 1957. Effect of crystalline electric fields on ferromagnetic anisotropy. *Phys. Rev.*, 108: 1152.

Wood, J.A., 1967. Chondrites: their metallic minerals, thermal histories and parent planets. *Icarus*, 6:1.

Yamamoto, N., 1968. The shift of the spin flip temperature of α-Fe_2O_3 fine particles. *J. Phys. Soc. Japan*, 24: 23.

NAME INDEX

Abe, H., 182
Ade-Hall, J.M., 129, 178
Aharoni, A., 93, 94, 178
Alperin, H.J., 183
Akimoto, S., 29, 33, 37, 38, 48, 91, 178, 181, 183
Alfvén, H., 174, 178
Amar, H., 59, 178
Anderson, P.W., 18, 85, 178
As, J.A., 136, 178

Bagina, O.L., 128, 178
Balsley, J.R., 79, 178
Banerjee, S.K., 23, 28, 29, 31, 32, 33, 34, 39, 85, 86, 93, 94, 99, 167, 168, 173, 174, 178, 180, 183, 184
Barr, K.G., 185
Bartholin, H., 31, 178
Bean, C.P., 70, 103, 104, 178, 181
Beetz, W., 128, 178
Belshé, J.C., 186
Bénard, J., 93, 178
Bertaut, F., 39, 178
Bhathal, R.S., 75, 78, 79, 108, 178
Bickford, L.R., 51, 178
Blackett, P.M.S., 162, 179
Blasse, G., 29, 179
Bömmel, H., 182
Bozorth, R.M., 37, 75, 179
Brailsford, F., 2, 179
Breiner, S., 160, 179
Brcun, J.A., 162
Brown, H.C., 79, 179
Brown, P.J., 183
Brown, W.F., 31, 45, 179
Brunhes, B., 162
Buddington, A.F., 79, 129, 178, 179
Bullard, E.C., 165, 179
Butler, R.F., 38, 63, 92, 173, 174, 179, 184

Callen, E.R., 23, 179
Callen, H.B., 23, 179

Carmichael, C.M., 92, 166, 179
Carmichael, I.S.E., 129, 179
Chaudron, G., 93, 182
Cheam, A.S., 184
Chevallier, R., 34, 86, 87, 88, 179
Chikhachev, V.A., 186
Cinader, G., 88, 179
Clarke, Mrs. S., vi
Cohen, J., 104, 179
Colburn, D.S., 184
Collinson, D.W., 122, 123, 128, 179
Colombo, U., 31, 179
Constabaris, G., 182
Cottrell, A.H., 56, 179
Cox, A.V., 137, 162, 163, 174, 179
Creer, K.M., 88, 103, 104, 136, 179

Dalrymple, G.B., 179
Day, R., 91, 180
De Boer, F., 93, 180
Dickson, G.O., 74, 110, 112, 180, 181
Dirac, P.A.M., 2
Doell, R.R., 137, 179
Domen, H., 168, 169, 180
Domenicali, C.A., 60, 180
Dunlop, D.J., 59, 61, 67, 70, 86, 99, 103, 111, 145, 180
Dzyaloshinsky, I., 35, 180

Endo, K., 88
Evans, M.E., 39, 70, 103, 180
Everitt, C.W.F., 40, 108, 180
Ewing, J., 128, 180

Fedoseeva, N.V., 186
Fitch, J.L., 145, 184
Flanders, P.J., 85, 91, 179, 180, 185
Fletcher, E.J., 23, 180
Folgheraiter, G., 162
Foster, J.H., 183
Frei, E.H., 178
Friedel, J., 56, 180

Frosch, A., 181
Fuller, M.D., 28, 36, 89, 114, 154, 180, 182, 186

Gazzarrini, F., 31, 33, 179, 180
Gibb, T.C., 178
Gillingham, D.E.W., vi, 113, 114, 118, 143, 144, 145, 178, 180
Girdler, R.W., 186
Goldstein, J.A., 171, 180
Goldstein, M., 185
Goodenough, J.B., 18, 180
Gorter, E.W., 168, 180
Graham, J.W., 75, 76, 79, 155, 165, 180, 181, 185
Graham, K.W.T., 186
Green, R., 156, 181
Greenwood, N.N., 178
Griffiths, D.H., 127, 181
Grommé, C.S., 129, 181
Gus'kova, Ye.G., 172, 173, 181

Haayman, P.W., 25, 186
Haigh, G., 85, 86, 133, 181
Hales, A.L., 186
Hamilton, N., 127, 181
Hanůs, V., 39, 181
Hargraves, R.B., 39, 173, 174, 178, 181
Harrison, C.G.A., 127, 181
Hayashi, I., 182
Hedley, I.G., 134, 181
Heezen, B.C., 183
Heirtzler, J.R., 165, 181
Heitler, W., 10, 181
Heisenberg, W., 10, 12
Herron, E.M., 181
Higbie, J.W., vi
Hirooka, K., 181
Holz, A., 70, 181
Howell, L.G., 89, 181

Iida, S., 85, 182, 185
Ince, P., 184
Irving, E., vi, 127, 156, 164, 181
Isacks, B., 164, 181
Ishikawa, Y., 37, 166, 181
Ising, G., 75
Iwagana, K., 182

Jacobs, I.S., 70, 181
Jaeger, J.C., 182
Johnson, C.E., 31, 178, 179
Johnson, E.A., 121, 122, 123, 125, 181

Johnston, M.J.S., 147, 150, 160, 161, 181, 185
Joplin, G., 185

Kapitsa, S.P., 149, 181
Katsura, T., 178
Kawai, N., 36, 129, 181
Keil, K., 184
Kern, J.W., 139, 140, 147, 150, 155, 158, 181, 182
Kersten, M., 56
Khan, M.A., 179
Khramov, A.N., 122, 127, 182
King, R.F., 75, 78, 79, 123, 126, 127, 180, 182
Kinoshita, H., 149, 153, 154, 183
Kittel, C., 8, 45, 46, 59, 72, 182
Kobayashi, K., 28, 36, 131, 133, 134, 182
Koenigsberger, J.G., 115, 128, 182
Kovach, R.L., 160, 179
Kovalenko, G.V., 186
Kramers, H.A., 18, 182
Kropáček, V., 168, 182
Krs, M., 39, 178, 181
Kumagai, H., 85, 182
Kume, S., 181
Kundig, W., 86, 87, 88, 182
Kushiro, I., 94, 95, 182

Landau, L.D., 31
Langevin, P., 1
Lanzavecchia, F., 31, 33, 179, 180
Larson, E.E., 33, 183, 185
Le Pichon, X., 181
Lewis, R.R., 39, 182
Lifshitz, E., 31
Lindquist, R.H., 182
Lindsay, J., 185
Lindsley, D.H., 129, 179
Livingston, J.D., 103, 104, 178
Lovering, J.F., 171, 172, 182, 185
Lowrie, W., 114, 182

Major, A., 127, 181
Martinez, J.D., 181
Mason, B., 170, 182
Matthews, D.H., 165, 186
Matthieu, S., 86, 179
Matuyama, M., 162
McDougall, I., 162, 182
McElhinny, M.W., 39, 70, 103, 180
Mercanton, P.L., 162
Merrill, R.T., 27, 37, 115, 182
Merritt, F.R., 178
Michel, A., 93, 182

NAME INDEX

Minnibaev, R.A., 168, 182
Mizushima, K., 85, 182
Morin, F.J., 35
Moriya, T., 35, 182
Morrish, A.H., 9, 38, 59, 182, 183, 184, 186
Murphy, T., 181, 185
Myasnikov, V.S., 182

Nagata, T., vi, 37, 38, 91, 109, 116, 146, 154, 162, 166, 183
Nathans, R., 35, 183
Néel, L., vi, 7, 14, 18, 19, 20, 21, 34, 35, 56, 59, 79, 80, 85, 96, 100, 104, 138, 143, 162, 165, 168, 183
Nicholls, J., 129, 179
Nininger, R.C., 86, 184
Ninkovitch, D., 164, 165, 183

Ogilvie, R.E., 171, 180
Ohnaka, M., 149, 153, 183
Oliver, J., 181
Ono, F., 181, 182
Opdyke, N.D., 165, 183
O'Reilly, W., 28, 33, 34, 167, 168, 178, 180, 183, 184
Osborn, J.A., 43, 44, 183
Ozima, Minoru, 27, 32, 33, 110, 183
Ozima, Mituko, 32, 110, 183

Pappis, J., 178
Parry, L.G., vi, 27, 57, 66, 68, 74, 75, 80, 81, 82, 83, 91, 92, 111, 112, 114, 115, 116, 125, 145, 167, 180, 182, 184, 185, 186
Patton, B.J., 145, 184
Pauthenet, R., 25, 85, 179, 183, 184
Peck, D.L., 181
Petrova, G.N., 182
Pickart, S.J., 183
Pitman, W.C., 181
Porath, H., 31, 85, 184
Pucher, R., 133, 184

Rado, G.T., 178
Raleigh, C.B., 31, 85, 184
Readman, P.W., 30, 33, 34, 183, 184
Rees, A.I., 75, 123, 127, 180, 182, 184
Remeika, J.P., 85, 178, 180
Riste, T., 25, 26, 184
Robson, G.R., 185
Romeijn, F.C., 186
Roquet, J., 112, 184
Runcorn, S.K., 179
Rusakov, O.M., 125, 126, 184

Sakamoto, N., 33, 183, 184
Sasajima, S., 181
Sawatsky, G.A., 186
Schieber, M., 178
Schroeer, D., 86, 184
Schuele, W.J., 91, 180
Schulkes, J.A., 168, 180
Schult, A., 21, 168, 184
Schwartz, E.J., 40, 184
Schwartz, K., 184
Searle, C.W., 38, 184
Selwood, P.W., 93, 180
Senftle, F.E., 39, 182
Shamsi, S., 154, 160, 184
Shimada, J., 182
Shimizu, Y., 101, 184
Shive, P.N., 38, 63, 92, 109, 184
Shtrikman, S., 179
Shull, C.G., 14, 184
Sironi, G., 179
Smart, J., 18, 184
Smit, J., 19, 22, 184
Smith, P.J., 162, 178, 184
Soffel, H.C., 59, 184
Sonnet, C.P., 174, 184
Srivastava, K., 179
Stacey, F.D., vi, 50, 51, 57, 60, 62, 67, 69, 71, 73, 75, 77, 78, 79, 85, 97, 99, 108, 109, 110, 116, 117, 122, 124, 125, 143, 144, 147, 150, 154, 155, 156, 157, 158, 160, 161, 165, 172, 173, 174, 178, 179, 180, 181, 184, 185
Statham, E.H., 181
Stephenson, A., 29, 185
Stoner, E.C., 43, 89, 185
Stopes-Roe, H.V., vi, 186
Stott, P.M., 155, 156, 157, 158, 185
Strangway, D.W., 63, 83, 91, 185
Strausser, W.A., 184
Stull, J.L., 178
Suhl, H., 178
Sunagawa, I., 85, 185
Sykes, L.R., 181
Syono, Y., 29, 48, 51, 166, 181, 185

Tarling, D.H., 162, 182
Tasaki, A., 85, 185
Tenzer, L., 25, 26, 184
Thellier, E., 116, 185
Thellier, O., 116, 185
Torreson, O.W., 162, 181, 185
Tropin, Yu.D., 99, 185
Tuck, G.J., 45

Uyeda, S., 63, 91, 120, 165, 166, 183, 185, 186

Van der Woude, F., 25, 26, 186
Van Vleck, J.H., 1, 13, 22, 186
Van Zijl, J.S.V., 164, 186
Verhoogen, J., 34, 110, 167, 168, 186
Verwey, E.J.W., 25, 27, 186
Vine, F.J., 165, 186
Vlasov, A.Ya., 91, 99, 134, 185, 186

Walsh, D.E., 179
Wasilawski, P.J., 135, 186
Watkins, N.D., 164, 168, 186
Weaving, B., 172, 173, 174, 186
Weiss, P., 4, 186
West, G.F., 145, 180
Westcott-Lewis, M.F., 91, 92, 167, 186
Whitworth, R.W., vi, 186
Wijn, H.P.J., 19, 22, 184

Williams, A.J., 179
Wilson, E., 155, 186
Wilson, R.L., 93, 163, 164, 168, 178, 186
Wise, K.N., 57, 62, 67, 69, 185
Wohlfarth, E.P., 89, 185
Wolf, W.P., 26, 186
Wollan, E.O., 184
Wood, J.A., 170, 174, 186
Wright, A.E., 181
Wright, T.L., 181

Yager, W.A., 178
Yamamoto, N., 86, 186
Yoshida, M., 178
Young, W.M., 39, 181
Yu, S.P., 59, 183

Zijderveld, J.D.A., 136, 178

SUBJECT INDEX

Activation energy, 32, 93, 100
—, electrical conduction, 26, 28, 32
Additivity of partial thermoremanences, 116–118
Ages, radiometric, v
Alternating field demagnetization, 70, 114, 120, 125–126, 134, 136–141, 154, 165, 169, 172
Aluminium substitution, 93
Angular momentum, 2, 3, 13, 26, 53
— — of Sun, 174
Anhysteretic Remanent Magnetization (ARM), vi, 126, 136, 141–145, 153, 169
Anisotropy, v, 22–23, 31, 35–36, 38, 40, 45, 63, 69, 72, 75–79, 85–86, 87, 89–91, 92, 118–120, 126, 130, 140, 146, 150, 158
— constants, 23, 28, 29, 31, 36, 38, 40, 46, 47, 48, 88, 89, 90, 154, 158
— energy, 47–49, 52, 54, 55, 59, 72, 87, 147, 148, 151–152, 155
— field, 48, 90, 91
— of ions, 34
Annealing, 57, 66, 67, 68, 86
Antiferromagnetism, 10, 12, 14–19, 35, 37, 39, 40, 88, 104
Antisymmetric wave functions, 11–13
ARM, see Anhysteric magnetization
Asteroids, 170
Atomic nature of magnetism, 1, 2, 4, 5–10
— orbits, 1, 2

Baked contact rocks, 162, 164, 169
Ball-milling, 33
Barkhausen discreteness, 60–62, 99
— effect, 60, 61
Basal plane of hematite, 85, 87, 88, 89, 91, 106, 130, 131
Basalts, 28, 31, 45, 84, 91, 92, 113, 117, 129, 156, 157, 158, 164, 168, 169
Blocking temperature, 33, 96, 97, 102, 105, 106, 107, 108, 114, 116, 118, 130
— volume, 130
Body-centred cubic lattice, 4, 14

Bohr magneton, 2, 25, 58
Brillouin function, 6
Brownian motion, 123–125, 126, 127
Burgers vector, 57

Canted antiferromagnetism, 19, 35, 38, 86, 87, 104
Carbonaceous chondrites, 170, 173, 174
Carbon steel, 98
Cassiterite, 39, 168
Cation-deficient compositions, 32, 33
Cation distribution, 8, 30, 31–33
C.G.S. (electromagnetic) units, vi, 1
Chemical remanent magnetization (CRM), 31, 34, 35, 84, 92, 94, 98, 105, 122, 128–135, 136, 137, 158, 172
Chondrites, 170, 172–175
Chondrules, 170
Clausius-Clapeyron equation, 94
Cleaning, magnetic, 27, 136–141
Closure domains, 46, 56, 110
Cobalt–copper alloy, 131, 133
Coercive force (H_c), 3, 28, 29, 34, 36, 37, 38, 56, 57, 58, 66, 67, 68, 69, 70, 71, 72, 73, 74, 79–83, 85–86, 87, 88, 91, 92, 101, 102, 103, 105, 106, 108, 109, 116, 125, 131, 132, 137–138, 139, 145, 170, 175
Coercivity of remanence (H_{cR}), 3, 70, 79–83, 86, 136, 138, 139, 141
Colloidal patterns, 36
Compensation temperature, 21, 38
Conduction, electrical, 26, 28, 30, 32
Contact zones, 163, 164
Continental drift, v
Cooperative phenomena, 5
Core, liquid, 174
Cosmic radiation, 175
Critical size, single domains, 59–60, 61, 103, 110, 133
— stress, 150, 155
Crystal defects, 35, 36, 55–56, 57, 61, 62, 67, 68, 69, 85, 86, 125, 175
Curie constant, 7, 20

SUBJECT INDEX

Curie point (θ_C), 4, 5, 7, 9, 10, 14, 17, 20, 28, 31, 33, 34, 40, 48, 59, 77, 88, 93, 94, 108, 116, 128, 129, 154, 160, 168, 172
— —, paramagnetic, 7
Curie's law, 6, 16
Curie-Weiss law, 7, 16, 17

Demagnetization by alternating fields, 70, 114, 125–126, 134, 136–141, 154, 165, 169, 172
— by cooling, 27, 36, 115
Demagnetized state, 5, 60
Demagnetizing factor (N), 43–44, 45, 64, 69, 71, 75, 79, 84, 103, 109, 139, 144, 149
Depositional Remanent Magnetization (DRM), 121–127
Detrital Remanent Magnetization (DRM), 121–127, 128
Diamagnetism, v, 1
Differential thermal analysis (DTA), 94, 95
Diffusion, ionic, 167, 168, 170, 172
Dipolar anisotropy constant, 36
Dipole field, 164
Direction cosines, 47
Dislocations, 55, 56, 57, 58, 62, 69, 92, 175
Dispersed grains, v, 3
Dolerite, 77, 156
Domain rotation, 49, 69, 70, 72, 74, 78, 79, 87, 115, 146, 147, 148
— theory, 1, 23, 41–65, 67–75, 105
— wall energy, 42, 52, 53–58, 59, 61, 64
— — moments, 60, 62
— — nucleation, 103, 133
— — pinning, 61, 72–74
— — thickness, 54, 57, 59, 62, 96, 110
— — translation, 56, 67–68, 72–73, 74, 78, 79, 99, 107, 108, 115
— walls, 9, 14, 41, 42, 52, 53–58, 61, 62, 67, 71, 72, 73, 74, 99, 105, 107, 110, 115
DRM, *see* Detrital Remanent Magnetization
Dykes, 45, 83–84, 163
Dynamo action, 174

Earthquakes, 158–160
Easy directions of magnetization, 22, 26, 48, 63, 72, 75, 78, 89, 106, 107, 120, 150
Eddy currents, 98
Electrical conductivity, 26, 28, 30
Electromagnetic units, vi, 1
Electron hopping, 25, 26, 28, 29, 32
— spin, 2
Ellipsoidal grain, 42–44
Epoch, geomagnetic, 163

Events, geomagnetic polarity, 163
Exchange energy, 13, 41, 53–54, 55, 59
— integrals, 12–13, 14
— interactions, 3, 4, 9, 10–14, 22, 53, 88, 134, 166
Exsolution, 34, 37, 39, 56, 63, 92, 129, 134, 158, 172

Fabric of rocks, v, 79, 89
Face-centred cubic lattice, 4
Felspars, 129
Fermions, 11
Ferric ions, 13, 19, 22, 25–35, 38, 39, 91, 92, 93, 168
Ferrimagnetism, 1, 4, 10, 12, 14, 19–21, 25, 40, 168
Ferrites, 19, 21
Ferromagnetic domains, 1, 4, 20, 41–65
— resonance, 31, 35, 85, 87
Ferromagnetism, v, 1–14, 41–65
Ferrous ions, 13, 19, 22, 25–35, 38, 39, 92, 99, 168
Field energy, 41, 69, 72, 73, 77, 105, 108
Fine structure anisotropy constant, 36
Flux density, magnetic (B), 1
Foliated rock, 89
Frequency factor, 96

Gabbro (S. Africa), 70, 103
Gallium substitution, 93
Geomagnetic reversals, 162–165, 169
Goethite, 39, 40
Gyromagnetic ratio, 2

Hard directions of magnetization, 48, 50, 69
— magnetic materials, 3, 4, 36, 80, 82, 89, 125, 136, 140
Helium gas, 93
Helmholtz coils, 75, 76
Hematite, 8, 19, 30, 31, 35–36, 37, 85–95, 106, 122, 125, 128, 129, 130, 131, 132, 133, 135, 136, 137, 158, 166
Hemo-ilmenites, 37, 98
Hexagonal axis (hematite), 35
— — (pyrrhotite), 39, 40
— close-packed lattice, 4, 87
History, magnetic, 4
Homopolar chemical bond, 12
Hund's rule, 13, 18, 19
Hydrogen substitution, 93, 94
Hyperfine magnetic fields, 25
Hysteresis, 3, 79, 86, 89–91, 136–138, 139, 142–143

SUBJECT INDEX 193

Igneous rocks, 35, 37, 105, 121, 129, 136, 155, 156
Ilmenite, 32, 37, 56, 63, 91, 129, 158, 166, 168
Ilmeno-hematites, 37, 98, 166, 167
Impurity defects, 35, 36, 85, 99, 167
Inclination error, 89, 126–127
Intergrowths, lamellar, 32, 63–65, 92, 129, 166
Intermolecular field, 4, 5, 6
− −, coefficient, 5
Internal field, 43, 44, 70, 73, 79, 80, 99–100, 101, 138, 139, 142
− stresses, 36, 85, 86, 88, 130, 132, 154
Interstitial lattice sites, 98
Ionic diffusion, 167, 168, 170, 172
− ordering, 30, 38, 166–167
Iron meteorites, 170–172
Irreversibility of magnetization, 3
Ising problem, 4
Isothermal Remanent Magnetization (IRM), 136, 154, 156, 173, 175
Isotopes, radioactive, v
Isotropic point of magnetite, 26, 27, 28

Kamacite, 170–172
Kilauea volcano, 129
Koenigsberger ratios, 115–116

L-type ferrimagnetism, 20–21
Lambda-point anomaly, 9, 10, 19
Lamellar domains, 45, 67, 70
− intergrowths, 32, 63–65, 92, 129, 166
Langevin paramagnetism, 102, 104, 123, 124, 131, 176
Lanthanum substitution, 93
Laterite, 93
Lattice defects, 35
− mismatch, 37
− parameters, 33, 37, 86, 132
− strain, 37
− vacancies, 30–33, 39, 40, 93, 94
− waves, 58
Legendre polynomials, 22
Lightning strikes, 136, 137
Low-temperature demagnetization, 27, 36, 115
− − oxidation, 30, 31, 32, 92, 128–130
− − properties, 8, 25, 31

M.K.S. units, vi, 1, 2
Maghemite, 30–31, 92–95
Maghemite–hematite transition, 92–95

Magnetic anomalies (ocean floor), 165
− cleaning, 27, 136–141
− colloids, 36
− dipole interaction, 10
− memory, 27, 36
− surveys, v
− viscosity, 96, 98–102
Magnetite, v, 8, 19, 21, 25–28, 31, 36, 41, 46, 47, 52, 56, 57, 59, 62, 63, 64, 66–84, 88, 91, 92, 93, 96, 101, 103, 106, 109, 110, 111, 113, 114, 115, 121, 122, 125, 128, 129, 130, 133, 134, 135, 138, 143, 144, 145, 148, 149, 154, 170
Magnetocrystalline anisotropy, 22, 26, 29, 31, 35, 36, 38, 40, 47–49, 63, 72, 77, 85, 89, 92, 103, 130, 146, 150
Magnetoelastic energy, 42, 49–52, 55, 63, 147, 148, 151–152, 155
Magnetomechanical factor, 2
Magnetostatic energy, 41–47, 56, 59, 63, 64, 77, 103, 108
Magnetostriction, 22–23, 36, 49–52, 57, 85, 86, 92, 98, 106, 146, 147, 148, 154, 155, 158
− and paleomagnetism, 155–158
− constants, 28, 29, 48, 146, 158
Magnetostrictive strain, 38, 55, 57
− − energy, 42, 49–52, 55, 63, 147, 148, 151–152, 155
Magnons (spin waves), 8–9, 14, 23, 58
Manganous oxide, 18–19
Memory effect, 27, 36
Metallic iron, 39
Meteorites, 39, 170–175
Molecular field, 4, 5, 6, 9, 10, 15, 18, 20, 21
− − coefficient, 5, 15, 18, 20, 21
Montecatini-Edison Laboratory, 30
Morin transition, 35, 36, 40, 86, 88
Mossbauer effect, 25, 26, 31, 35, 86
Multidomain grains, 28, 58–60, 62, 64, 65, 67, 69, 70, 71, 72, 84, 90, 99, 105, 107–109, 110, 113, 114, 115, 116, 118, 119, 125, 130, 133, 134, 138, 139, 144, 145
− thermoremanence, 27

N-type ferrimagnetism, 20–21, 38, 168
Natural remanent magnetization (NRM), 31, 92, 101, 115, 121, 125, 129, 155, 162, 168, 169, 172, 173, 174
Nebula, solar, 174
Néel construction, 79–80, 138, 142, 143
− point, 14, 15, 16, 17, 20, 35, 37, 39, 40, 129

Néel theory (ferrimagnetism), 19–21, 25, 29
Neutron diffraction, 14, 25, 35
Nickel, thermoremanence in, 109
Nickel-iron, 170–172
Neutron irradiation, 175
Non-dipole field, 164
Nuclear magnetic resonance (NMR), 2
Nucleation of domain walls, 103, 133

Ocean-floor spreading, 164–165
Ocean ridges, 164–165
– trenches, 164
Octahedrally coordinated ions, 21, 25, 27, 30, 35, 36, 168
Orbital magnetic moment, 22, 23, 25
– wave functions, 11–13
Order, short/long range, 5, 7, 9, 38
Order–disorder transitions, 9, 14, 19, 30, 38, 40, 92, 98, 167, 168
Orthohombic magnetite, 25, 26, 27
Oxidation of magnetite, 30–31, 92, 93, 128, 129, 134
Oxidized basalt, 31, 91, 129, 158, 164
– titanomagnetites, 32–35, 56, 63, 134–135, 167
Oxygen ions, 18, 27, 30, 33, 36, 40

P-type ferrimagnetism, 20–21, 29
Paleo-intensity determinations, 129, 133
Paleomagnetic poles, vi, 129
– stability, v, 70, 82, 98–102, 116, 130, 154, 175
Paleomagnetism and magnetostriction, 155, 158
Paramagnetism, v, 1, 6, 7, 14, 102, 104
Parent bodies of meteorites, 174
Partial anhysteretic remanence (PARM), 145
– thermoremanent magnetizations, 97, 116–118, 134, 145
Pauli exclusion principle, 13, 18
Periodic table, 13
Permeability, 1
– of free space (μ_0), 1, 2
Permeameter, 71
Perturbation theory, 12–13
Phase transition, in nickel-iron, 172
Phonons (lattice waves), 58
Piezomagnetic effect, v, 24, 146–161
Piezoremanent magnetization (PRM), 147, 154
Planck's constant, 2
Poisson's ratio, 52
Polarization, magnetic, 3
Porphyry, 156
P.s.d. moments, see Pseudo-single domain effects

Pseudo-brookite, 129
Pseudo-dipolar model, 23
Pseudo-single domain (p.s.d.) effects, 27, 60–62, 74, 92, 106, 107, 109, 110–114, 115, 120, 125, 133, 139–141, 144, 145
Pyroxenes, 70, 103
Pyrrhotite, 39, 40

Quantum mechanics, 1, 2, 4
Quenching, orbital moments, 23, 26
–, high temperature phases, 36, 40, 129

Radiation damage, 175
Radioactivity, v
Rationalized M.K.S. units, vi, 1, 2
Red clay, 31
– sandstone, 31, 40, 128
Reduction of hematite, 133, 135
Reversal of remanence, 29, 37, 38, 40, 91, 92, 134, 162–169
Rhombohedral axis (hematite), 35, 87
– phase, 91
Richardson-Einstein-De Haas effect, 2, 3
Rigidity, 52
Rod-shaped domains, 46
Rotation of domains, 49, 69, 70, 72, 74, 78, 79, 87, 115, 146, 147, 148
Rotational hysteresis, 89–91

Saturation magnetization (I_S), 3, 4, 28, 31, 48, 59, 64, 76, 77, 86, 93, 104, 131, 168
– remanence (I_{RS}), v, 3, 59, 79–82, 101, 102, 103, 105, 109, 125, 132, 144, 145
Sea-floor spreading, 164–165
Secular variation, 121
Sedimentation, 121–125
Sedimentary rocks, 35, 37, 75, 105, 121, 125, 128, 129, 136, 163, 164
Seismomagnetic effect, v, 154, 158, 161
Self-demagnetizing factor, 43, 44, 45, 71, 75, 79, 84, 109, 139, 144, 149
– – field, 43–45, 70, 71, 83, 108, 120, 138, 143
Self-reversal of magnetization, 29, 37, 38, 40, 91, 92, 134, 162–169
Shales, 40
Shape anisotropy, 45, 69, 70, 72, 76, 85, 103, 130
– – of dyke or sill, 83–84
Sills, 45, 83–84
Single domains, 27, 42, 58–60, 61, 62, 63, 65, 69, 70, 71, 72, 89, 96, 98–99, 102, 103,

SUBJECT INDEX 195

104, 105–107, 110, 113, 114, 116, 130, 131, 132, 133, 134, 139, 145
Soft magnetic materials, 3, 73, 74, 80, 136
Solar magnetic field, 174–175
– nebula, 174
Specific heat, 9–10, 19
Spinel structures, 21, 30, 31, 93
Spin-flop transition, 35
Spin magnetic moments, 2, 3, 10, 22, 23
Spin-orbit coupling, 22, 47
Spin-wave functions, 11, 12
– waves, 8–9, 14, 23, 58, 96
Spontaneous magnetization, 3–10, 14, 19, 20, 21, 23, 25, 26, 28, 29, 30, 31, 33, 35, 38, 64, 77, 85, 86, 87, 89, 91, 92, 93, 104, 106, 108, 145, 158, 166
Stability of magnetization, v, 70, 82, 98–102, 116, 130, 135, 154, 175
Stacking faults, 31, 35
Stoke's Law, 123
Stony-iron meteorites, 170–175
Stress, effect on magnetization, 50, 106, 146–161
– sensitivity of remanence, 150, 153, 154, 159, 160
– – of susceptibility, 149–150, 154, 159, 160
Sub-lattice moments, 34, 40, 168
Sub-lattices, 7, 16, 19, 20, 21, 26, 35, 36, 104, 168
Sulphides, 39
Superantiferromagnetism, 88, 102, 104
Superexchange interaction, 18, 19, 22, 25, 38, 165, 167
Superparamagnetism, 33, 88, 96, 102–104, 105, 106, 107, 130, 133
Surveys, magnetic, v
Susceptibility, 1, 5, 20, 45, 49, 70–75, 80, 81, 82, 83, 84, 87–88, 89, 103, 104, 105, 107, 118, 133, 140, 142, 143, 144, 146–150, 156, 157
–, paramagnetic, 7, 14, 15
Symmetric wave functions, 11, 12

T-Tauri phase of sun, 174–175
Taenite, 170–172
Tasmanian dolerite, 77
Tectonomagnetism, 150, 158–161
Ternary ferrites, 168
Tertiary ferrites, 168

Tetrahedrally coordinated ions, 21, 25, 27, 93, 168
Thermal activation, 32, 58, 96–104, 107
– agitation, 4, 25, 123, 145
– demagnetization, 117, 125, 128, 172, 174
Thermoremanent magnetization (TRM), vi, 27, 31, 40, 62, 64, 65, 70, 79, 84, 91, 92, 96, 97, 105–120, 121, 125, 129, 130, 132, 134, 135, 136, 137, 139–140, 145, 153, 154, 155, 156, 158, 163, 167, 169, 172, 173, 175
Tin-substituted hematite, 36, 168
Titanium ions, 28–30, 33, 38, 92
– substitution, 28–30, 36, 37–38, 91–92
Titanohematite, 32, 33, 35, 37–38, 89, 91
Titanomaghemites, 33, 135, 168
Titanomagnetite, 21, 22, 23, 28–30, 33, 51, 55, 56, 59, 69, 70, 71, 77, 91, 99, 128, 129, 135, 136, 146, 149, 150, 151, 154, 155, 166, 167, 168
Topotactic process, 30
Torque magnetometry, 35, 75, 76, 77, 85, 89
Transition (iron series) elements, 13, 22
Trenches, ocean, 164
Trigonal axis (hematite), 35, 87
TRM, see Thermoremanent magnetization
Troilite, 39, 170
Twinning, 36, 85

Ultrabasic rocks, 168
Ulvöspinel, 28, 30, 129

Valence electrons, 13
Van Vleck's rule (scalar multiplication of spin vectors), 13
Van Leeuwen's Theorem, 1
Varved clay, 121, 127
Vector model of atom, 2
Viscosity coefficient, 100, 102
– of water, 122
Viscous magnetization, 136, 137
Volcano-magnetic effect, 154, 158–161
Water-flow effects, 126–127
Wave functions of orbital electrons, 10–12
Weak ferromagnetism, 35–36, 88
Weathering, 31
Weiss theory (quantum-corrected), 4
Widmanstätten structure, 171, 172, 174

X-ray diffraction, 31, 33

Young's modulus, 52